James A. Tyner
Academic Writing for Geographers

De Gruyter Contemporary Social Sciences

—

Volume 29

James A. Tyner

Academic Writing for Geographers

——

A Handbook

DE GRUYTER

ISBN 978-3-11-162932-2
e-ISBN (PDF) 978-3-11-118972-7
e-ISBN (EPUB) 978-3-11-119054-9
ISSN 2747-5689
e-ISSN 2747-5697

Library of Congress Control Number: 2023938940

Bibliographic information published by the Deutsche Nationalbibliothek
The Deutsche Nationalbibliothek lists this publication in the Deutsche Nationalbibliografie;
detailed bibliographic data are available on the internet at http://dnb.dnb.de.

www.degruyter.com

Acknowledgments

In Lewis Carroll's *Alice's Adventures in Wonderland*, the titular character follows a White Rabbit down a rabbit hole, never once considering how she will get out again. *Academic Writing for Geographers* is my White Rabbit, and for over 15 years I have explored the rabbit's warren of the written world. As an academic, I am well aware of the anxieties, stress, confusion, and frustration that surround academic writing; but I am familiar, also, with the joys that come with it. Since I first began this journey in 2008 it has been my hope to help graduate students and early-career faculty navigate the Wonderland of academic writing.

I am deeply appreciative of Andre Borges and the entire editorial team at De Gruyter for their support of this project, from our initial conversations to the final product. This book, quite literally, would not exist were it not for their willingness to take a chance on a book such as this. During my long journey I have encountered also many scholars—editors and reviewers—who have provided positive and constructive criticisms on earlier iterations of my manuscript. Many remain anonymous but my gratitude does not.

It is important to have a support network of friends and colleagues who can provide encouragement, wisdom, and guidance. As an undergraduate and, later, graduate student, I was fortunate to encounter many scholars who helped me along the way: Stuart Aitken, Bernie Bauer, Larry Brown, Keith Collins, Jeff Crump, Michael Dear, Mona Domosh, Larry Ford, Ernst Griffin, Audrey Kobayashi, John O'Leary, Gary Peters, Laura Pulido, Curt Roseman, Doug Sherman, and Rod Steiner. Sadly, many of these individuals have passed, but their influence remains.

For a book on academic writing, I want to acknowledge the many students and colleagues with whom I've collaborated on commentaries, encyclopedia entries, articles, book chapters, and books:

Derek Alderman, Gabriela Brindis Alvarez, Sutapa Chattopadhyay, Chhunly Chhay, Jaerin Chung, Corrine Coakley, Alex Colucci, Gordon Cromley, Andrew Curtis, Christabel Devadoss, Dan Donaldson, Mindy Farmer, Chenjian Fu, Sam Henkin, Donna Houston, Josh Inwood, Sokvisal Kimsroy, Rob Kruse, Olaf Kuhlke, Jay Lee, Hanieh Haji Molana, Mandy Munro-Stasiuk, Erik Parker, Mark Rhodes, Stian Rice, Andrew Shears, Eric Sheppard, Savina Sirik, Zheye Wang, Rachel Will, Xinyue Ye, and Huanyang Zhao. In ways both big and small I have gained invaluable insight on academic writing from these scholars, and I express my appreciation to each and every one.

From kindergarten onward—if not earlier—I wanted to be a writer. On birthdays and holidays I would 'write' stories in 3x5-inch spiral-bound notebooks and give these as gifts to my relatives; and I would tell stories to my brother who, in

https://doi.org/10.1515/9783111189727-001

turn, would draw pictures to illustrate my imaginings. It is to my family, my parents Gerald and Judith, my brother, David, and my aunt, Karen that I express my most heartful thanks. Over the decades they have encouraged my dreams of becoming an author, and I thank them for their support. Closer to home, literally, I thank my daughters, Jessica and Anica Lyn; and my life's partner, Belinda, for their support of my writing. Belinda, in particular, has been my guiding light; her love for our family and her support for my passion illuminate even the darkest days of the often solitary life of a writer. Thanks also to Carter and Bubba, our seven-year-old and four-year-old rescue dog and cat, respectively, who truly show me each day what is important in life: a soft bed to sleep in. Lastly, though, I return to give thanks to my mother, Judith. An accomplished author in her own right, Mom has been with me since Day One in my writing life. When I was in kindergarten, it was Mom who typed my first 'story', and since that time we have talked at length all-things writing. We shared our hopes, dreams, and frustrations, often over a cup of hot chocolate at the local bookstore; and we read and commented on our respective papers and books. And so, it is only fitting that I dedicate this book, on writing, to Mom.

Contents

Preface

I published my first 'scholarly' article in 1993. It did not come easy and, looking back, I'm somewhat embarrassed by what I wrote. But write it I did, and I'm still proud of that first piece. Also, I emphasize 'scholarly' because I had written material previously. In high school I worked on the school newspaper, mostly writing articles for the sports page but also, occasionally, writing other pieces. I also dabbled with poetry and published my first and only poem when I was an undergraduate student at California State University at Long Beach. Looking back, I've been lucky, I suppose. I always wanted to write. Since before I could write, I wanted to be an author. I wanted to follow in the footsteps of J.R.R. Tolkien and Edgar Rice Burroughs; I wanted to write fantasy stories of elves and orcs and follow my characters on epic quests in heroic struggles against evil. I compiled lists—many lists—of characters; I named them and described them and gave them purpose. I drew maps of their worlds and planned routes they would travel.

And I read: mostly fantasy novels, of course. Beyond Tolkien and Burroughs, I especially enjoyed the *Dragonlance Chronicles* by Margaret Weis and Tracy Hickman, and the *Sword of Shannara* trilogy by Terry Brooks. But I read also many of the so-called canons of literature, namely Ernest Hemingway, John Steinbeck, Joseph Conrad, Samuel Beckett, William Faulkner, and Lewis Carroll. I was drawn especially to the early dystopian fictions of H.G. Wells, Aldous Huxley, and George Orwell. And, as an aspiring author, I also read books on writing in general and on writing fantasy in particular. In the 1980s and 1990s, *Writer's Digest* published a series of books on "The Elements of Fiction" and I spent many long hours reading and re-reading Orson Scott Card's *Characters and Viewpoint* and Monica Wood's *Description*. I pored over Ann Hood's *Creating Character Emotions*, Marc McCutcheon's *Building Believable Characters*, Dwight Swain's *Creating Characters*, and Tom Chiarella's *Writing Dialogue*. I reveled also in Card's *How to Write Science Fiction and Fantasy* and J.N. Williamson's edited volume, *How to Write Tales of Horror, Fantasy, and Science Fiction*. These tomes still occupy a privileged position on my bookshelves.[1]

---

1 J.N. Williamson (ed.), *How to Write Tales of Horror, Fantasy, and Science Fiction* (Cincinnati, OH: Writer's Digest Books, 1987); Orson Scott Card, *Characters and Viewpoint* (Cincinnati, OH: Writer's Digest Books, 1988); Orson Scott Card, *How to Write Science Fiction and Fantasy* (Cincinnati, OH: Writer's Digest Books, 1990); Dwight V. Swain, *Creating Characters: How to Build Story People* (Cincinnati, OH: Writer's Digest Books, 1990); Jack M. Bickham, *Scene and Structure* (Cincinnati, OH: Writer's Digest Books, 1993); Monica Wood, *Description* (Cincinnati, OH: Writer's Digest Books, 1995); Marc McCutcheon, *The Writer's Digest Sourcebook for Building Believable Characters* (Cincin-

https://doi.org/10.1515/9783111189727-002

My desire to write fantasy novels, in part, is what led me to geography. In 1985 I enrolled in physical geography, my first ever geography course, taught by Gary Peters. I thought that learning about weather and climate and mountains and vegetation would help add an element of realism to my stories. After taking that first introductory course on physical geography, I found a new passion—or so I thought. I was smitten by geography. When I had enrolled at CSULB the previous year, I planned to major in English, with an emphasis on creative writing. During my first three semesters, I took courses on poetry and short-story writing; I still regret not taking the class on novel writing. But after passing Gary's class—I think I received a 'B'—I wanted to pursue a career in geography. That I was starting anew is an understatement. Both in middle school and high school I mostly enrolled in English courses. But, like any good hobbit, I persevered. I no longer enrolled in writing courses and instead attended classes on geomorphology and soils and the environment. I supplemented these courses with elective classes in geology and oceanography. It wasn't until I took classes at the University of Bradford, in England, that I switched my area of interest to more human concerns.

History also was a favorite subject of mine. Tolkien's *The Hobbit* and his *Lord of the Rings* trilogy are histories of Middle Earth. I wanted to write my own histories, of important characters in their own geographies. To add a degree of realism I sought out historical accounts of 'everyday' life, such as Joseph and Frances Gies' *Life in a Medieval City* and Paddy Griffith's *The Viking Art of War*.[2] I supplemented these with books on writing historical fiction, notably Sherrilyn Kenyon's *Everyday Life in the Middle Ages: The British Isles from 500 to 1500*.[3] I didn't realize it at the time, but I was, albeit haphazardly, doing the research required for any form of writing, fiction or non-fiction.

I think, in retrospect, my career as an academic writer owes much to the insight and inspiration I drew from Card, Wood, McCutcheon, and so many others. Not that I can conjure fictional dragons. No, what I gained was something more tangible, more practical. For what these books did—and continue to do—was to help me understand how to *read* academic journals and books and, in turn, how to *write* academic journals and books. It is important to learn by doing; but it is equally important to learn what others are doing. By reading these 'how-to' books, I learned early on that, unlike Gandalf, it isn't possible to wave a magic

nati, OH: Writer's Digest Books, 1996); Tom Chiarella, *Writing Dialogue* (Cincinnati, OH: Story Press, 1998); and Ann Hood, *Creating Character Emotions* (Cincinnati, OH: Story Press, 1998).
2 Joseph Gies and Frances Gies, *Life in a Medieval City* (New York: Harper Perennial, 1981 [1969]); and Paddy Griffith, *The Viking Art of War* (London: Greenhill Books, 1995).
3 Sherrilyn Kenyon, *Everyday Life in the Middle Ages: The British Isles from 500 to 1500* (Cincinnati, OH: Writer's Digest Books, 1995).

wand and see a book emerge fully formed. Writing was a skill, something that one learned both through trial-and-error and practice. In some ways, I was becoming a character in my story: I began to learn the craft of writing, not unlike how fantasy characters learned to use sorcery or a sword.

There were no book-length treatments of geographic writing, however. Indeed, as I was to discover, many (most?) geography departments were (are?) woefully negligent in preparing undergraduate and graduate students to learn how to write. Methods courses were plentiful: statistics, cartography, geographic information systems, remote sensing.[4] As Dydia DeLyser observed more than two decades ago, writing appeared to be "something that we 'just do', hoping it will turn out well."[5] This was certainly my experience. As a student many years ago, there was little material to help me learn how to write *as a geographer*. Certainly, there was nothing comparable to the bounty of books on writing characters or dialogue. I could—and did—turn to more generic writing manuals and guides. And these did provide some help. What I really wanted, though, was a book that addressed writing within my own discipline of geography. *Academic Writing for Geographers* is a response to that need.

My objectives are straightforward: to provide a guide on writing abstracts, book reviews, encyclopedia entries and commentaries, journal articles and book chapters, and scholarly monographs specific to the discipline of geography. More broadly, I present a brief guidebook that is oriented and readily accessible to senior undergraduate and graduate students and early-career faculty who feel intimidated by, or otherwise unclear about, the process of writing scholarly pieces for publication.

I should also say something about what this book is not. In his book on writing science fiction and fantasy, Card understood that a book, especially a brief introduction, cannot "tell you everything you need to know about writing."[6] My foray into this subject is no different. I do not provide chapters on writing theses or dissertations; nor do I include chapters on writing blogs, newspaper editorials, and the like. For better or for worse, these are not peer reviewed. Why does this matter? It matters because, as graduate students or early-career faculty—and especially those who are precariously employed—our jobs are often dependent on peer-reviewed publications. For many academics, DeLyser explains, "writing often forms an important part of our work, one rendered visible (and dreadedly countable) in

---

4 When I was a student, there were no qualitative courses offered in my departments.
5 Dydia DeLyser, "Teaching Graduate Students to Write: A Seminar for Thesis and Dissertation Writers," *Journal of Geography in Higher Education* 27, no. 2 (2003): 169–181; at 170.
6 Card, *How to Write*, 1.

'outputs' yet seldom exposed in its private practice."[7] In other words, academic writing is made public—and evaluated—primarily in refereed publications; in turn, these publications are frequently weighed and used to evaluate our 'productivity'. For those geographers on the tenure track, for example, both tenure and promotion guidelines habitually include any number of quantitative criteria. Many departments, for example, require a specific number of refereed journal articles for tenure and promotion; additional criteria might specify that these articles appear in top-tier journals with high impact factors (i. e., journal rankings). I know of several departments that have elaborate formulas to 'objectively' assess one's scholarly contributions: a solo-author, refereed article in a top-tier journal earns the candidate five points; a co-authored, refereed article in a second-tier journal earns two points; an editorial in the local newspaper receives zero points; and so on. We may disagree with this approach; but our disagreement does not negate the reality that, for many of us, we frequently 'write' in a stressful environment "of escalating demands and 'metric dogma' pursued by university managers."[8] Consequently, as Rae Dufty-Jones and Chris Gibson find, "pressures brought to bear on the production and value of research writing are jeopardizing personal and community relationships and enflaming a crisis of mental and physical health in the academy."[9] DeLyser echoes this concern: writing becomes nothing more than 'academic labor' and, as such, "we can easily slip into a loss accounting where 'writing time' must be 'carved out' of other hours in the day—the labor of writing must take us away from life's other important labors, both paid and unpaid." She concludes that "in overtaxed, neoliberalized, highly metricized, competitive, and increasingly precarious academic careers, it often feels this way. ... And that can lead to the prevalent emotions of fear and anxiety related specifically to writing (particularly among emerging and earlier-career scholars)—a pervasive problem, one that may make us feel isolated as individuals and even mired in failure."[10] Simply put, many aspirant academics are overworked, struggle with self-doubt, and harbor feelings that they don't belong. In a word, they feel like imposters. What should be an enjoyable part of our profession becomes burdensome and we become alienated from writing. Compounding the problem, many academics

---

7 Dydia DeLyser, "Writing's Intimate Spatialities: Drawing Ourselves *to* our Writing in Self-Caring Practices of Love," *EPA: Economy and Space* 54, no. 2 (2022): 405–412; at 406.
8 Rae Dufty-Jones and Chris Gibson, "Making Space to Write 'Care-fully': Engaged Responses to the Institutional Politics of Research Writing," *Progress in Human Geography* 46, no. 2 (2022): 339–358; at 339.
9 Dufty-Jones and Gibson, "Making Space to Write," 339–340.
10 DeLyser, "Writing's Intimate Spatialities," 407.

—especially women and scholars of color—work in toxic environments.[11] As Julia Zielke and co-authors identify, in an ideal world, for those working at a university, "one's institutional home ... should provide a nurturing, protective, caring environment that allows for curiosity-driven exploration, personal growth, and development of an ethical, reflective disposition of the world around us."[12] Such toxic geographies cannot but negatively affect one's writing.

Writing doesn't have to be—and shouldn't be—alienating. I embrace fully De-Lyser's optimism that it is possible to rethink writing as a "loving practice ... as something we want to make space for in our lives."[13] However, I am sensitive to the fact that some academics may never 'love writing' but simply need to just 'get through it.' This is a point raised by a reviewer of an earlier draft of this manuscript. It is important to respond both to the needs of aspiring writers who want to perfect their craft and those academics who, for whatever reasons, need to meet the writing demands expected of their profession. This, in a nutshell, is why I cover those writing conventions that often prove the most stressful.

Neither is *Academic Writing for Geographers* a procedural book on how to submit articles or book proposals. On the one hand, there are several existing sources that cover this ground. Often, these contributions are written by editors and from the vantage point of editors—best to get this information straight from the proverbial horse's mouth. On the other hand, these procedures are highly idiosyncratic. Beyond the specific formatting styles required, journals and book publishers have particular submission processes and it is best to consult these at the time of submission. This is particularly important in that changing technologies quickly render some procedures obsolete. Now, we upload files to portals like Manuscript Central. When I first submitted manuscripts, however, I *mailed* three copies with an accompanying 5¼-inch floppy disc. And many years ago, when I tried to publish poems, I typed the poems using carbon-copy paper. I must admit, there was something viscerally satisfying in sending off those envelopes.

Nor is this a book on organization. I say next to nothing on how to organize one's files or the best way to keep tabs on bibliographies. Again, there are many sources on organizational techniques and I encourage readers to consult these.

---

11 See for example Laura Pulido, "Reflections on a White Discipline," *The Professional Geographer* 54, no. 1 (2002): 42–49; Minelle Mahtani, "Toxic Geographies: Absences in Critical Race Thought and Practice in Social and Cultural Geography," *Social & Cultural Geography* 15, no. 4 (2014): 359–367; and Tianna Bruno and Cristina Faiver-Serna, "More Reflections on a White Discipline," *The Professional Geographer* 74, no. 1 (2022): 156–161.
12 Julia Zielke, Matthew Thompson, and Paul Hepburn, "On the (Im)possibilities of Being a Good Enough Researcher at a Neoliberal University," *Area* (2022), doi: 10.1111/area.12815.
13 DeLyser, "Writing's Intimate Spatialities," 410.

Lastly, this is not a book on research methodologies. Let's face it: a common source of frustration among many writers, from the novice graduate student to the seasoned professor, is not knowing *what* to write. I'm clueless on many subjects, for example climatology. Having a single course on the subject as an undergraduate in no way prepared me to compose a manuscript of any length on boundary layer climates or barotropic wave propagation. To contribute in any meaningful way on such topics requires *research*. I can't help you learn *what* to write; for that, you need to immerse yourself in your area of interest. I can help you learn *how* to write. This is not a subtle distinction and harkens back to the perennial laments expressed by DeLyser and others. As graduate students, we take seminars and courses, ostensibly to learn what to write—we learn to conduct research. Unfortunately, many of us don't have the opportunity to learn how to write. That's why I've written *Academic Writing for Geographers*. Whether you write on biogeography or biopolitics, fluvial landforms or memorial landscapes, it's my hope that you'll gain a better understanding of how to write as a geographer in an academic setting.

And now, having dispensed with the preliminaries, let's begin.

Chapter 1 Academic Writing in Geography

Some days I sit at the computer and nothing happens. I stare blankly at the computer screen—and the screen stares emptily back at me. My fingers hover above the keyboard, periodically touching down like gentle raindrops; but words don't form and sentences don't flow. Troubling thoughts cloud my mind: Am I kidding myself? Can I really write this manuscript?

This book is designed to help geographers write scholarly pieces in an academic setting. My focus is on the stalwarts of academic writing—journal articles, book chapters, books, book reviews, abstracts, encyclopedia entries, and commentaries. This is not to discount the importance of other formats and forums. Indeed, there is a tremendous need for geographers to engage in *public geographies*, for example in the dissemination of ideas via fiction, poetry, blog posts, newspaper op-eds, policy papers, and so on. As Rob Kitchen explains, these forms of *praxes* that extend beyond conventional academic papers and books can be used to create new public geographies that seek and enact progressive change.[1] Nevertheless, for many academics (and quite a few non-academics) conventional academic formats continue to assume tremendous importance. For geographers, the need to publish articles, book chapters, book reviews, and the like in scholarly journals will not disappear anytime soon. Often, these forms of writing also prove the most stressful for graduate students and early-career faculty.

Students and early-career academics often feel frustration because they are unsure how to write. They sit down at the computer to compose an abstract, or book review, or chapter, and draw a blank because they don't yet have the skills to craft the piece of writing required. Where can geographers turn for guidance? Most graduate departments in geography don't offer courses on writing; and while students may be encouraged to look elsewhere, such as English departments or writing centers, these options are imperfect because they are not attuned to the oddities of geography. And for early-career academics, or people on the job market, there are even fewer places to turn. Helpfully, many geographers have in recent years taken the initiative to form writing circles or writing clubs. But the sad fact remains: many aspiring geographers don't have many good options. And so they turn, understandably, to general guides to research writing, such as Kate

---

1 Rob Kitchen, "Engaging Publics: Writing as Praxes," *Cultural Geographies* 21, no. 1 (2014): 153–157; at 156. See also William G. Moseley, "Engaging the Public Imagination: Geographers in the Op-Ed Pages," *The Geographical Review* 100, no. 1 (2010): 109–121.

https://doi.org/10.1515/9783111189727-003

Turabian's *A Manual for Writers of Research Papers, Theses, and Dissertations.*[2] Others seek out more specialized books on academic writings, for example how to write 'qualitative' research articles. There is, however, no comprehensive book that covers the various forms in which practicing geographers are expected to write and publish. *Academic Writing for Geographers* addresses this need.

Toward an Epistemology of Writing

It is common for 'how-to' guides to offer well-meaning (but sometimes shallow) advice. Consider Donald Fiske's chapter on "Planning and Revising Research Reports."[3] In his section on "The Writing State," Fiske writes: "Before you decide what content you will include and (even more important) what content you will not be able to include, before you organize your paper, decide on the journal to which you will submit it."[4] This is sound advice. But what of the content? After selecting a particular journal, Fiske continues that "In your writing, conform to the style and format of your target journal." Again, good counsel. At this point, however, Fiske explains that in research articles "you have to persuade your peers about the quality and significance of your findings and interpretations." He suggests—and this is helpful—that "you can get some idea about how convincing your paper is by getting some colleagues to read it."[5] Okay; but wait. Where's the part about actually *writing the paper?* We seem to have jumped over the most important part, moving from targeting a journal to having colleagues read the draft manuscript. This is a pretty important step in the process, one that many students and even more seasoned academics find challenging. How do we begin a manuscript? How do we organize the manuscript? What sections do we include? In what order do we present our material? It is not enough to *describe* the writing process; we need to *analyze* the writing process. Basically, we need to be *knowledgeable* about writing, with knowledge here referring to the skills acquired by a person through experience or education.

Epistemology, Phil Hubbard and colleagues explain, "concerns how knowledge is derived or arrived at; assumptions about *how* we can know the world." They

2 Kate L. Turabian, *A Manual for Writers of Research Papers, Theses, and Dissertations*, 7th edition (Chicago: University of Chicago Press, 2003).
3 Donald W. Fiske, "Planning and Revising Research Reports," in *Writing and Publishing for Academic Authors*, 2nd edition, edited by Joseph M. Moxley and Todd Taylor (Lanham, MD: Rowman & Littlefield, 1997), 71–82.
4 Fiske, "Planning and Revising," 77–78.
5 Fiske, "Planning and Revising," 79.

elaborate: epistemology "concerns how we can validly come to know something."[6] To that end (and with apologies to my colleagues in philosophy who probably balk at my usage of the term), I propose an *epistemology* of geographic writing. In the following chapters, I evaluate written works with the purpose of highlighting the craft of writing. This is not to provide generic, step-by-step instructions on compiling notes or managing reference lists; it is not to bring clarity to the submission process or how to engage in collaborative writing. These are important elements and are well covered elsewhere. My writing of this short book, and of the particular approach I take, stems from years of disappointment as students and junior colleagues especially struggle to *understand* the craft of writing for scholarly publications.

Metaphors and analogies are helpful to convey one's ideas and so I'll begin with a story that, in slightly different form, is probably pretty familiar. Writing is a skill and to be an effective writer, it takes effort. My youngest daughter, Jessica, took ballet lessons. She didn't immediately go on stage, her feet bound in pointe shoes, and perform *pirouettes*. No, she began (as a five-year-old) jumping over cloth mats and stretching her limbs. Years of stretching and years of rote movements—until her body was effectively transformed into that of a ballerina. Only when Jessica's calf muscles had developed sufficiently was she permitted to begin training *on-pointe*. And before I scare off some readers, I'm not suggesting that it will take years before you're sufficiently skilled to *begin* writing, let alone actually writing. To be an effective writer, though, there is some truth to this. As a doctoral student I recall a conversation I had with Michael Dear. Michael was on my PhD committee and was a seasoned writer whom I respected. One day, after a seminar class—probably on postmodernism, but I don't really remember —I asked him if my writing was "good enough" to publish. At the time, I wasn't intimidated by writing. I had always wanted to be a writer; but this was different. Prior to enrolling in the graduate department at the University of Southern California I had spent many years thinking about writing and had in fact published *one* poem. Clearly I wasn't the next Robert Frost but—and this is the point—I wasn't anxious about writing. Without hesitation, Michael replied: "Probably in one to two years." Michael was direct; and for me, this was invaluable. My writing *was* rough and my thoughts on paper too jumbled. I knew what I wanted to say. By this time I had been working on my dissertation for two years but I was not yet skilled enough to effectively commit my ideas to paper. For my daughter, training translated into countless hours of learning conventional forms, such as the *arab-*

6 Phil Hubbard, Rob Kitchin, Brendan Bartley, and Duncan Fuller, *Thinking Geographically: Space, Theory and Contemporary Human Geography* (London: Continuum, 2002), 5.

esque and the *croisé*. In geography, we rarely receive such in-depth training on the craft of writing. Usually, we're just tossed into the deep end of the pool: sink or swim, publish or perish.

In hindsight, I was fortunate. "The emotions around academic writing," Dydia DeLyser explains, "can feel unpredictable, even overwhelming; entering into scholarly conversation with one's most-admired (and/or most dreaded) peers can provoke anxiety, fear, inadequacy, and imposter syndrome."[7] In my case, I didn't feel inadequate but I did know that I had much to learn. In return, I hope to now share some of that knowledge I've acquired over the years.

The Process of Writing

Writing, Dydia DeLyser suggests, is an inherently spatial process. By this she means that "academics who wish to be successful writers must master, typically with little or no training, four basic spatialities of writing: making the time to write, finding a place to write, and creating what can be called the 'head-and-heart space'—the conceptual-and-emotional space—in which to write."[8] In other words, one does not simply sit down to write. The act of writing is an intimate and emotional act.[9] It is also deeply personal—beyond the myriad social and institutional relations that circumscribe our daily lives. As such, I'm somewhat wary to provide guidelines (however well intentioned) on creating the spaces conducive to one's writing experience. What works for me won't necessarily work for someone else. And that's okay. What's important—I truly believe—is that you begin to know something about yourself through the writing process. Writing, after all, is an emotional experience. One moment I may feel euphoria, the next, despair. Still, despite the idiosyncrasies inherent to the writing process, I want to say *something* that hopefully will resonate with you.

So, let's start with some basic considerations: the where, when, and how of writing. Some people write in coffeehouses and others write in cozy nooks with scented candles and Beethoven playing in the background. Some people prefer writing in the morning; others write in the evening; and still others write whenever they (hopefully) find time. Some writers are obsessive in their routines: get

---

7 Dydia DeLyser, "Writing's Intimate Spatialities: Drawing Ourselves *to* our Writing in Self-Caring Practices of Love," *EPA: Economy and Space* 54, no. 2 (2022): 405–412; at 409.

8 DeLyser, "Writing's Intimate Spatialities," 408.

9 Jenny Cameron, Karen Nairn, and Jane Higgins, "Demystifying Academic Writing: Reflections on Emotions, Know-How and Academic Identity," *Journal of Geography in Higher Education* 33, no. 2 (2009): 269–284.

up every morning at 6am and write a minimum of 500 words.[10] Other writers struggle to maintain any semblance of order. My writing routine has changed over the years. When I was younger, I wrote most often in the evening. I seemed to think more clearly late at night. Perhaps it was the overall silence. Now, I find that I am most effective when writing in the early morning, usually from 7am until 11am. I am not obsessive; I don't set aside blocks of four solid hours every morning, although I do *try* to write almost every day. When I arrive at work I try to set writing as my first priority. This means I do not chat with colleagues and I don't obsess over emails. Certainly, if some other project is pressing, such as having to review tenure and promotion materials—I put off writing. Likewise, if students drop by the office, my door is always open and I'll stop writing to answer whatever problems they have. However, in the spirit of full transparency, rarely are students seen at seven in the morning.

So, what exactly do I do when I sit down to write? The easy answer is, it depends. At any given time I'm usually juggling several projects. Some authors I know use the 'assembly line' metaphor: one project in the planning stages, one project under active construction, and one project in production. In other words, I'm often planning a prospective manuscript, actively writing a manuscript, revising a manuscript, and perhaps editing or copy editing a manuscript. This appears too linear, I think, but it does capture the reality that I don't start a project and work on nothing else until that project is completed. I know some authors who do work that way and they are very successful in their careers. To underscore what will become a recurrent theme, what works for me may not work for someone else. With this in mind, let's say a few more words about the writing process. However, rather than providing a series of tasks—for example, planning, drafting, revising, and editing—I'll provide a couple of concrete examples.

Let's suppose you're planning an article that draws on critical race theory, the Black Lives Matter movement, and the policing of public space in Los Angeles. How do you start? In our example, you have at least three key concepts in play (social movements, policing, and public space); you also have a concrete (empirical) case study (Black Lives Matter) in a specific location. Do you start with general discussion of public space? Alternatively, do you begin with an overview of critical race theory? The options are seemingly endless. For many writers, this results in a paralyzing frustration.

---

10 Personally, I've always been wary of a too-structured routine. Over the years, I've encountered many students and faculty who became frustrated and riddled with guilt when, for example, instead of writing during their allotted time, they had to tend to sick children or a rambunctious puppy.

Or consider a project on the Khmer Rouge and the Cambodian genocide (1975–1979). I know that events that happened prior to 1975 are important. From 1970 to 1975, there was a devastating civil war that contributed to the rise of the Khmer Rouge. But, of course, a critical component of the civil war was the coup against the former head of state, Prince Norodom Sihanouk. This can lead, though, to a consideration of the Sihanouk regime, which leads to a consideration of the French colonial period, which leads to a consideration of pre-colonial Cambodia, and on and on it goes. There is also an important geography to the Cambodian genocide, in that the revolution that brought the Khmer Rouge to power was inseparable from the revolutionary movement(s) of neighboring Vietnam. And the revolutions in Vietnam were significantly impacted by the Cold War geopolitics that pitted the United States and other Western governments against the Soviet bloc. And yet, the Soviet bloc was not monolithic and, in fact, the Sino-Soviet split played a decisive role in the subsequent geopolitical relations between Cambodia and Vietnam. You get my point. Where do I begin my story and what do I include?

In Robert Frost's well-known poem, "The Road Not Taken," a traveler (Frost?) comes upon two paths that diverge in a forest. I find in Frost's poem—apart from its geographic metaphor—a useful guide to brainstorming about *beginnings*. For where do I go if I choose one path instead of another? How will my article proceed if, for example, I begin with critical race theory, followed by a discussion of public space? Where does my article go if I start with a vignette of the Black Lives Matter movement, before moving to an engagement with critical race theory? Frost is correct, in that the path chosen will make all the difference. However, and unlike the poet's traveler, we can easily turn around and try a different path—that is, we can *revise* our early attempts. I'll have more to say about this later. For now, I want to clarify that the purpose of the planning stage is to work through the many possibilities, with the understanding that the destination (the final, written product) is greatly impacted by the beginning.

So how do we begin to think about beginnings? Some writers utilize an approach known as *rapid free writing*. Here, the writer will—within a strictly limited time span—write down words and phrases about a topic. Some use white boards or chalk boards; others use legal pads of paper. The method is less important than the activity of simply getting ideas down. And here, I often think of the project in one of two ways. For some projects, notably books, I think of these projects as novels, with associated characters, plots, and scenes. Theories, concepts, and models comprise the characters; arguments (or theses) form the plot and sub-plots; and the contextual framing makes up the scenes.

When not thinking of my writing projects as stories I might view them as puzzles. Here, the planning of writing consists of figuring out, first, if I have all the pieces and, second, trying to fit the pieces together. Regardless of analogy, I begin by sketching a rough outline. If I'm approaching the project like a story, what is the *main* plot? If I'm approaching the project like a puzzle, what is the completed image? What is that plot or image? A piece of advice my PhD advisor, Curt Roseman, offered many years ago was that if you can't explain your thesis in under 50 words, you probably don't yet have a good grasp of the project. Of course, the point isn't to get hung up on 50 words. Curt didn't obsess over 48 or 52 words; his point was that I needed to be as concise as possible to effectively convey my argument or thesis in written form.

By the time I have a rough outline, I should have a pretty good idea where I'm heading. Let's say, for example, I'm writing an article and it has six (exceptionally generic) sections: Introduction, Theoretical Context, Study Site, Data and Methodology, Analysis, and Conclusions. The first thing I do is open a Word file, and list in bold face these six sections. I then—from my brainstorming activities—jot down key ideas or even fully fledged sentences that will appear in the relevant sections. I'm not worried about punctuation, grammar, or word choice. Frequently, I will include notes to myself, set apart in brackets, for example [*need to clarify term*] or [*make connection with competing concepts*]. The point is to begin to add flesh to the skeleton.

When I write, whether I'm drafting an article, a book chapter, or a book, I don't always start at the beginning. In fact, I rarely do. I may work on the theory section first; the next day, perhaps I develop the data/methodology section. In other words, I work on different sections, at different times, and rarely in sequential order. Frankly, on any given day, some sections (for various reasons) seem more straightforward to write than others. In general, however, the introduction and conclusions are held off till the end. Only after I have a fairly good sense of the direction of the manuscript do I begin to draft the introduction. Certainly, I have an *idea* of where the paper will go; and this has to be expressed in the introduction. Other writers approach this stage very differently—a point raised by several reviewers of this manuscript. As one reviewer explained: "I *always* start with the first sentence (and do not start until I have decided on that sentence, even if I know it is likely to change in the revision) and write straight through until I get to the end (even if I am unsatisfied with the development or logic in some parts.[11]

Feelings of doubt often manifest themselves during the drafting of the manuscript. The key, as Jenny Cameron and her co-authors explain, "is not to erase difficult emotions from writing but to find in them their productive potential rather

---

[11] I'm especially grateful of the comments provided by this and several other reviewers who shared their insights into their writing processes.

than paralysis."[12] When I write, I often experience a range of emotions throughout the process. Sometimes I feel exhilarated; other times, I want to throw the entire project into the waste-basket (or, more appropriately, delete the entire file). Over the years, I've come to recognize a *pattern* to this emotional roller-coaster. Now, when I feel frustrated, I step back and ask myself, *Why am I feeling frustrated? What is the problem?* Often, it's because the puzzle pieces don't fit together. So, I return to my guiding thesis and ask again of the manuscript, *What am I trying to say? Why does this section not fit? It is too tangential? Does this concept warrant its own section?*

Often, my feelings of frustration are grounded in something concrete about my writing. Consequently, these frustrations may serve as internal editors, perhaps signaling that something doesn't quite fit. Usually, after a few stressful days, I'm able to work out the kinks in my thinking and continue on. It's because of years of enduring these frustrations that I've come to work with instead of against these emotional roller-coasters.

The actual revision of a manuscript does not necessarily begin when the manuscript is fully drafted. Often, writers proceed iteratively, writing and revising as they go along. When I edit, I begin with the nuts-and-bolts: tightening sentences and checking for proper grammar, spelling, punctuation, diction, and all the other bits-and-bobs that make for effective, readable writing. However, my primary concern is the coherence of the manuscript. This is just as important for the humble abstract as it is for book-length monographs. Is the order of material organized effectively? Are the key concepts, arguments, or conclusions clearly articulated? In my 2018 monograph, *The Politics of Lists*, I drew upon several concepts—for example, archives, lists, networks, surveillance practices—and the most challenging aspect was determining how best to present these concepts. I also think about the many stylistic conventions to liven up the manuscript. This doesn't mean that similes pop up like dandelions in every paragraph. It does mean that I think about the cadence of a sentence or the effective use of alliteration and assonance. Relatedly, tone is important. I write often of violence, especially genocidal violence. A manuscript on torture and execution should not be written in a light-hearted style; there is no room for whimsy when conveying the pain, the suffering, or the anguish that comes with such atrocities. And to those who wonder why I raise this point, I'll just mention that I once reviewed (and rejected) a manuscript on genocide that was full of puns.

The process of revising a manuscript actually appears in two different guises: a private face and a public face. The private guise refers to the revisions we make

---

12 Cameron et al., "Demystifying Academic Writing," 274.

as we write and re-write the manuscript. Most of these revisions take place unseen; these are the edits I make before I share the draft manuscript with anyone else. Once I cross the Rubicon and ask, for example, a colleague to read over the manuscript, my writing begins to assume a more public persona. At this point, I'm subjecting my writing to the evaluative eyes of others, opening myself for (hopefully) constructive criticism. And when I finally submit the manuscript to an editor, my work truly assumes a public face. Now, editors and external reviewers will make comments and criticisms. Not all feedback will be positive. Some, in fact, can be quite negative. This imparts an additional emotional component to the writing process as we are forced to negotiate the sometimes treacherous waters of hostile reviews. Once submitted, a decision ultimately will be rendered. At one end of the spectrum, the manuscript may be accepted with minor or major revisions; at the other end, the manuscript might be rejected. Most commonly, decisions will land somewhere in between, for example a 'revise-and-resubmit.' This is okay. It indicates that reviewers and editors see promise in your work and, with sufficient re-working of the material, the manuscript stands a good chance of acceptance.

What if the manuscript is rejected?

Having a manuscript rejected is *not* pleasant. It can be frustrating, infuriating, depressing; it can be all of these, plus myriad other emotions. However, there is *always* hope. If a manuscript is rejected, you need to read closely the reviews and the editorial decision. Why was the manuscript rejected? Perhaps you targeted the wrong journal. This is actually an easy correction to make. Resubmit the manuscript elsewhere. Perhaps the reviewers gauged that the paper tried to do too little or, alternatively, too much. Here, you need to *read* the reviews and evaluate their comments and criticisms closely, keeping in mind your overall purpose. Let's face it. Not all reviews are good reviews. By this, I mean that not all reviewers are correct in their evaluation of your manuscript. As most seasoned writers can tell you, there are instances where reviewers simply missed the point. If this happens, contact the editor and clarify the mis-readings. Above all, read critically and respond concretely to the reviews. Don't automatically concede to the demands of the reviewers; often, these comments and criticisms are meant as helpful suggestions —but they are not always helpful. Sometimes, external reviewers *read into* the manuscript and want you to take the manuscript down a different path. You need to decide if that alternative course is appropriate or if you should stick to your original course.[13]

---

13 By way of example, my initial draft proposal of this section did not include a sustained discussion of the review process. I thank Don Mitchell for his constructive suggestion.

In the end, the process of writing constitutes a series of interrelated tasks, alternating between reading and writing, but reading and writing *in different ways* throughout the *different phases* of preparing a manuscript. Every manuscript tells a story; and there is a story behind the writing of every story. In both instances, you're the writer. And this necessitates a final consideration after having authored a manuscript, whether the end-product is an abstract or encyclopedia entry, a book or a book chapter.

Some Final Comments about Academic Writing in Geography

Let's be frank. Academic writing, including academic writing in geography, has a bad reputation.[14] Deservedly or not, geographers are not known for their prose. In part, this perception stems from the fact that much of our training downplays the importance of writing.

As graduate students, we are required to demonstrate our scholarly credentials through the writing of a dissertation; the finished product appears nothing more than a means to an end. In our post-graduate lives, journal articles (especially) become utilitarian, the literal building blocks of one's academic career. In turn, for many researchers, there is little enjoyment in writing; the task becomes a chore, something that detracts from more pleasurable pursuits. DeLyser effectively conveys this attitude:

> As students and former students, most of us were probably trained in the 'writing it up' school of writing, where writing is a simple, even almost mechanical act that comes (often hurriedly) at the very end of research in order to link that research (the data we gathered and created) with the analysis we perform in our offices or labs and in our brains.[15]

More broadly, as Katherine Burlingame finds, "there is a growing concern for the proliferation of disenchantment in research and in writing that has become increasingly inaccessible, dull, and mechanical."[16]

Writing does not have to be turgid and the writing process does not have to be robotic. Academic writing does not have to be written in academese. Academic

---

14 See for example Robert M. Wilson, "Writing Geography: Teaching Research Writing and Storytelling in the Discipline," *EPA: Economy and Space* 54, no. 7 (2022): 1450–1459.

15 Dydia DeLyser, "Writing Qualitative Geography," in *The SAGE Handbook of Qualitative Geography*, edited by Dydia DeLyser, Steve Herbert, Stuart Aitken, Mike Crang, and Linda McDowell (Thousand Oaks, CA: SAGE, 2010), 341–358; at342.

16 Katherine Burlingame, "Where are the Storytellers? A Quest to (Re)enchant Geography through Writing as Method," *Journal of Geography in Higher Education* 43, no. 1 (2019): 56–70; at 57.

writing does not have to be pompous, and it certainly does not have to be dull. This doesn't mean that everything we write should flow in iambic pentameter. Nor should we punctuate our papers with parables and personification—although alliteration is always allowed! Simply put, writing should communicate our ideas —clearly and effectively. And on this point, good communication is as important for the geographer who studies aeolian sediment transport as it is for the geographer who studies environmental racism.[17]

Teaching academic writing in geography is challenging. So too is writing about academic writing in geography. The problem stems from geography's eclectic persona. Geography straddles several disciplines, each with their own conventions and traditions that impact what *and how* we write. When I entered the master's program in geography at San Diego State, my classmates included anthropologists, meteorologists, geomorphologists, biogeographers, and population geographers (as I then considered myself). The problem wasn't simply the so-called 'split' between human geographers and physical geographers; it wasn't that some 'human' geographers were ignorant of albedos and surface friction or that 'physical' geographers scoffed at any project lacking a hypothesis. And it certainly wasn't the (artificial) divide between qualitative and quantitative geographers. Instead, the problem was that we approached our subjects conforming—albeit unknowingly—to conventions long held and long promoted in our areas of specialization. We did not yet have the skills or the experience to recognize and appreciate how these habits affected our abilities to communicate with each other, especially in written form. Ultimately, I hope that *Academic Writing in Geography* can talk across the divides. I write not as a human geographer, a population geographer, a political geographer, or any of the other 'labels' that attempt to pigeon-hole my work. I write—and I encourage you to write—as a geographer.

---

17 In the course of this book I draw exclusively on examples of what I consider effective writing samples. I don't discount the fact that we can learn from 'bad' writing, that is, poorly structured manuscripts or jargon-laden texts. There is much to learn by knowing what *not* to do and, frankly, there are abundant examples. That said, I'm reluctant to call out poor writing. Social media—and even a few commentaries published in academic venues—is rife with snarky comments, for example mocking run-on sentences or incessantly long paragraphs. But ridicule is not useful. Indeed, a frequent complaint of novice writers is fear of public ridicule or derision of their writing. Why, then, would I deliberately call out examples of writing I consider less successful? I want to encourage writers to read effectively, but this means reading compassionately.

Chapter 2 Abstracts

An abstract, simply stated, is a concise distillation of a document. As John Fraser Hart explains, "the abstract should not be confused with an introduction, which places the article in a larger context of knowledge, nor with a summary, which is a longer and more detailed restatement of procedures, findings, and conclusions."[1] Instead, an abstract conveys the *essence* of a presentation, article, book chapter, or book. "A good abstract," Hart says, "is like a bouillon cube, not like the wrapper; it should present the true essence of the article, not merely list its ingredients."[2]

When advising students, I encourage them to "read widely" and to "read often." On the one hand, this advice helps students gain a broad understanding of the discipline and, on the other hand, contributes to their writing skills. This advice though, and however well intentioned, is not without its problems. Time constraints, coupled with the voluminous amount of research published, inhibit our ability to read thoroughly. From child-care to elder-care to self-care, we are confronted in our day-to-day lives with innumerable choices. This impacts our ability to actually read widely and read often and, unfortunately, can add to a sense of frustration. Choices have to be made. We must become finicky in our research, putting off some books and articles in favor of others that are more immediately relevant to our current tasks. Abstracts facilitate the academic triage many of us must practice.

Paradoxically, many novice writers (and even more seasoned academics) don't always appreciate the importance of the abstract. James Wheeler laments, "to many writers the preparation of an abstract is an unwanted chore required at the last minute by an editor or insisted upon even before the paper has been written by a deadline-bedeviled program chair[person]."[3] And in full disclosure, I sometimes find myself guilty of such a charge. Frequently, I arrive at work, eager to upload a manuscript—only to be halted by the dreaded 'submit abstract' step in the process. Wheeler, of course, is correct in his lamentation and in his cautionary remarks, for "perfunctory, last-moment abstracts" often "do not live up to the quality of the manuscript itself, so that readers may unwisely choose not to read the article because of the 'inadequate abstract.'"[4]

---

1 John Fraser Hart, "Ruminations of a Dyspeptic Ex-Editor," *The Professional Geographer* 28, no. 3 (1976): 225–232; at 227.
2 Hart, "Ruminations," 227.
3 James O. Wheeler, "Writing Abstracts," *Urban Geographer* 17, no. 4 (1996): 283–285; at 284.
4 Wheeler, "Writing Abstracts," 283.

https://doi.org/10.1515/9783111189727-004

Abstracts are most commonly encountered in academic journals, but they appear elsewhere, notably in in conference programs and edited books. Regardless of where they are encountered, all abstracts perform the same general functions.[5] First, the abstract enables the reader to gauge whether the manuscript (or presentation) is of sufficient relevance to warrant reading (or attending). Second, abstracts may help overcome language barriers and facilitate the dissemination of research in multilingual settings. Increasingly, journals provide abstracts in multiple languages, notably Mandarin Chinese, Spanish, and French but, depending on the regional scope of the journal, other translated abstracts may be provided. Thus, while the main text is written in the parent language, readers of other languages are able to access the central themes of the manuscript. Third, the abstract can foreshadow important concepts, themes, or methods by using key words that are subsequently used in the main text. Fourth, the abstract can facilitate comprehension of the key arguments of the text.

Given that the purpose of an abstract is to communicate effectively and efficiently the details of completed research, it is advisable that abstracts are written toward the end-stage of the writing process. In reality, we are often compelled to write abstracts before the research has been completed. Indeed, it is ironic that abstracts frequently are hurriedly written *after* a project but other abstracts often are hurriedly written *before* a project. For this reason, I prefer to classify abstracts into two main forms, these being *preliminary abstracts* and *end-product abstracts*. Preliminary abstracts are condensations of research written prior to the completion of the proposed presentation or manuscript. There are five common types of preliminary abstracts: (1) participation in conferences; (2) call for papers; (3) invited conference/workshop presentation; (4) invited contributions of special edited journals; and (5) invited contributions to special edited books. These are not mutually exclusive and will be discussed in turn.

Prior to participation in conferences, prospective presenters are commonly asked to submit an abstract, usually of 150–300 words, detailing the material they propose to present. In an ideal world, the research has been completed and the presenter can simply craft an abstract based on that work. However, conference registration is often many months in advance and presenters submit their abstracts *in anticipation* of completing the research prior to the conference. In this scenario, the presenter is compelled to speculate on the finished product. As detailed later, this significantly mediates the material that can be included in an abstract.

---

5 Cate Cross and Charles Oppenheim, "A Genre Analysis of Scientific Abstracts," *Journal of Documentation* 62, no. 4 (2006): 428–446; at 429.

Abstracts written in response to a circulated 'call for papers'—often abbreviated as 'cfp'—constitute a second form of preliminary abstract. Here, the researcher is made aware of a particular event, such as an organized conference. Here, the organizer (or organizers) has drafted a brief essay that details the purpose of the event. Key questions or concepts may be raised for prospective presenters or authors to consider, and it is not uncommon for specific works (e.g., exemplary articles or books) to be referenced. Preliminary abstracts, therefore, must be oriented toward the explicit goals and key themes and concepts forwarded in the call for papers.

A third form of preliminary abstract is the invited request to participate in a conference or workshop. In this case, the organizer(s) extends a personal invitation to a scholar (or a team of scholars) and requests their participation in a particular conference or workshop organized around a particular theme or topic. The invitation (usually) includes a brief summary—itself an abstract!—of the proposed event and, possibly, some general questions or subjects to address.

Finally, two related forms of preliminary abstracts are those written in response to invitations to contribute to special edited journals or books. As with invitations to participate in a particular conference or workshop, scholars are expressly invited to contribute an article for a special issue of a journal or a chapter for an edited volume. Notably, when invited to submit an abstract, whether for a conference/workshop, special issue, or edited book, the invited scholar generally has more leeway in writing their proposed abstract and the organizers are more likely to offer constructive criticism to ensure that the proposed contribution conforms with the overall project.

End-product abstracts are those that convey the essence of a finished presentation or manuscript. Most often, end-product abstracts are written for journal articles submitted for publication. In addition, for special edited journals or books, preliminary abstracts are converted into end-product abstracts, in that the authors are able on completion of the article or chapter to revise the former abstract to reflect the completed manuscript. Rarely are end-product abstracts written for conference presentations and it is almost unheard of that preliminary abstracts for conference participation are transformed into end-product abstracts *for that conference*. On this last point, presenters may attempt to convert their conference presentation into a manuscript for possible publication.

Abstracts as Miniature Manuscripts

Effective abstracts convey the essence of a proposed research project (in the case of preliminary abstracts) or of a completed project (in the case of end-product abstracts). Regardless, all abstracts share some common elements. Chiefly, abstracts convey something about why the research was conducted; how the research was conducted; and what the research demonstrates. Beyond these three elements, and depending on the word space available, additional information can be included. Rather than simply recanting a list of do's and don'ts, let's reverse engineer some abstracts. My focus will be end-product abstracts, as these are the abstracts that (usually) are reviewed and published, but let's start with one example of a preliminary abstract.

In anticipation of the 2019 conference of the American Association of Geographers, Joshua Inwood and Derek Alderman submitted the following preliminary abstract:

Long remembered and venerated as the first Capitol of the Confederacy and more recently for George Wallace's vitriolic support of segregation, over the last decade the City of Montgomery, Alabama has worked to reinvent itself. Montgomery officials and private developers have invested in its downtown core and worked to redevelop the downtown business district and more recently have declared the city to be the 'capital of cool.' Interspersed among the new restaurants, baseball stadium, boutiques and high rise apartments are reminders of the city's past and its connection to some of the most divisive moments in US history. This paper focuses on the urban core's contested memorial landscape, efforts at urban redevelopment, and contradictory visions of Montgomery's connection to its past. Specifically, we argue Montgomery, Alabama's development vision attempts to navigate the raw activism of 'memory work' and the taking of responsibility for history's racism, but is also indicative of how Montgomery and other Southern cities appropriate commemorative change to serve the growing market influence of civil rights tourism, neoliberal urban development and promotion, and strong opinions about the valorization of white supremacy in Southern cities. As a result, we can use efforts at redevelopment in Montgomery to understand the use and appropriation of memory and the ways cities in the Southeastern United States are envisioning urban development in the context of broader changes to our understandings of how race operates in the United States. This paper is based on fieldwork completed in the summer of 2018 and portions of this project are funded through a National Science Foundation grant. It utilizes qualitative research including open-ended interviews and archival research.[6]

As a preliminary abstract, this is pretty informative. The location of the research is articulated, reference is made to methodological procedures, and several key concepts are introduced. Stylistically, the abstract is written (mostly) in the first-person, active voice; Inwood and Alderman repeatedly use phrases such as "we argue." Note also that the abstract is fairly abstract in specific findings; this follows because the abstract, in anticipation of the conference, was written and submitted well before the manuscript was finalized. Clearly, the abstract (and future presen-

---

6 This unpublished abstract is used with permission.

tation) is based on *completed* research; however, the precise presentation was still being planned. In fact, as Inwood explains, the abstract and eventual paper came about through numerous conversations during the course of their field work. As Inwood recalls,

> During our downtime we were spending a lot of time at a local coffee shop located within the Kress Building and in exploring the memorial landscape of the city. The abstract here was sketched out on napkins and in a notebook at the coffee shop as we began to think about what we were observing in the city. The paper that ultimately developed out of these conservations over hot chocolate and coffee was different than what we created here.[7]

Inwood goes on to explain that the paper that was presented at the conference, and the subsequent journal article, were both considerably different from the ideas sketched in this preliminary abstract. According to Inwood, the core argument remained, with the substantive change being a more focused case study of the Kress Building and proximate redevelopment activities in the city.

Inwood and Alderman's example is important also because it illustrates that preliminary abstracts are often *works-in-progress*, in that they reveal a particular moment in time during the course of one's research. Accordingly, as the writing continues, so too will preliminary abstracts transform into end-product abstracts. By way of closure, here's Inwood and Alderman's end-product abstract as it appears in published form:

> Scholars are increasingly studying memory-work as an essential place-defining force within cities, but few scholars have analyzed urban redevelopers as agents of memory- work. Using the Montgomery Builds effort to redevelop the Kress Building as a 'memory moment,' we argue for a broader reading of memory-work that recognizes the broad spectrum of social actors, interests, and tensions involved in not only doing justice to the legacies of racialized pasts but also appropriating them in the service of urban capital. Central to our argument is a recognition that urban spaces are not just the product of the labor of remembering and preserving, but that these spaces have an affective and material place and impact within people's lives and connections with the past. In so doing, we articulate how memory works through the remaking of space and place and argue for a broader definition of memory-work, a recognition of the harder and softer socio-political forms they can take in cities, and the way ostensibly painful memories are folded back into urban redevelopment visions in ways that facilitate but also complicate development and racial reconciliation.[8]

---

7 Joshua Inwood, personal correspondence, April 19, 2021.
8 Joshua Inwood and Derek H. Alderman, "Urban Redevelopment as Soft Memory-Work in Montgomery, Alabama," *Journal of Urban Affairs* (2020), doi: 10.1080/07352166.2020.1718507.

Here, Inwood and Alderman write in the first-person, active voice. Some material present in the preliminary abstract is noticeably missing, for example the fairly generic and unhelpful statement, "qualitative research including open-ended interviews and archival research." The end-product abstract is tighter, cleaner, and provides a greater level of *conceptual* specificity. Indeed, the emphasis shifts in the two abstracts, from a compilation of several competing ideas to a more focused engagement with memory-work. In short, end-product abstracts, unlike preliminary abstracts, are written in tandem with completed manuscripts. To that end, abstracts should be direct and complete, free from speculative sentences, such as 'We hope to find ....' There is no hoping. By the time one writes an end-product abstract, both the research and the manuscript are completed. The abstract should reflect that.

In the *Journal of Geophysical Research: Atmospheres*, Scott Sheridan and Cameron Lee published an article titled "Temporal Trends in Absolute and Relative Extreme Temperature Events Across North America."[9] Interestingly, this journal requires *two* abstracts: the conventional abstract and a 'plain language summary.' According to the journal's home page, the plain language summary "should be written for a broad audience that includes journalists and the science-interested public. It should state the general problem, describe what research was conducted, the result, and why the findings are important. The plain language summary improves article discoverability and allows your paper to be more widely accessible."[10] Helpfully, the inclusion of both conventional abstracts and plain language summaries allows us to highlight key features of abstracts and how these differ from other renditions, such as 'plain language summaries.'

Here's Sheridan and Lee's conventional abstract:

In this research, we define extreme temperature events using a recently defined excess heat factor, based on the exceedance of apparent temperature beyond the 95th percentile along with an acclimatization factor, to define extreme heat events (EHE). We extend the calculation to assess cold and develop relative metrics to complement the absolute metrics, where extremeness is based on conditions relative to season. We thus examine extreme cold events (ECE), relative extreme heat events, and relative extreme cold events in addition to EHE. We present a climatology of these variables for North America, followed by analyses of trends from 1980 to 2016. While EHE and ECE are found in the core of summer and winter, respec-

---

9 Scott C. Sheridan and Cameron C. Lee, "Temporal Trends in Absolute and Relative Extreme Temperature Events Across North America," *Journal of Geophysical Research: Atmospheres* 123 (2018): 11,889–11,898.

10 American Geographical Union, "Checklist for Submitting a Paper to an AGU Journal," available at https://www.agu.org/Publish-with-AGU/Publish/Author-Resources/New-Manuscript-Checklist (accessed April 13, 2021).

tively, relative events tend to have a broader seasonality. Trends in relative extreme heat events and EHE are upward, and relative extreme cold events and ECE are downward; the relative events are changing more rapidly than the absolute events.[11]

At just 153 words, Sheridan and Lee's abstract conveys considerable information. The immediate purpose is to define *extreme heat events* based on a calculation of *extreme temperature events*. This is extended to develop *relative* metrics for both cold and hot weather events. Specific measurements are based on data compiled for North America between 1980 and 2016. Findings indicate how extreme heat and cold events are changing, and how relative events are changing with respect to absolute events. Superficially, the abstract appears to lack some information that an outside reader might need to understand the research. My meteorology background, for example, is limited to a single course taken as an undergraduate. When reading Sheridan and Lee's conventional abstract, I can understand what they're doing and why, up to a point. Concretely, however, I'm at a loss. Given my (lack of) training in meteorology, several concepts are undefined. Absolute and relative events? I have no clue. Does this suggest that the abstract is poorly written? No—because I'm not the target audience for the conventional abstract. Let's see what Sheridan and Lee say when writing for a broad audience that includes non-specialists, such as myself.

> One of the most critical ways in which weather conditions influence the environment is through extreme temperature events. While excessive heat and cold conditions have been amply studied, events that are extreme relative to the time of the year have been less examined. These relative events may grab fewer headlines but can have important impacts on the environment, agriculture, and human health. In this research, we present a climatology of cold and heat events, both absolute and relative, for North America, followed by an analysis of how they have changed from 1980 to 2016. Results show an increase in heat events and decrease in cold events across most of the United States and Canada. More interestingly, the relative events are changing slightly more rapidly than the absolute events.[12]

In the plain language summary, Sheridan and Lee write in an informative but informal tone. They explain that extreme temperature events are vitally important, especially as these relate to the environment, agriculture, and human health. However, according to Sheridan and Lee, extreme events are understudied, especially when considered *relative* to the time of the year. This is something many of us have experienced: during the winter months, for example, when temperatures average around 20°F, a day with 60°F temperatures seems really hot. To that point, Sheridan

---

11 Sheridan and Lee, "Temporal Trends," 11,889.
12 Sheridan and Lee, "Temporal Trends," 11,889.

and Lee analyze heat waves and cold waves both in an absolute setting and a relative setting, and conclude that relative events are changing more rapidly than absolute events.

In an ideal world, it would be helpful if all end-product abstracts were accompanied with plain language summaries. Until then, however, most of us will continue to write conventional abstracts and thus have to negotiate between writing for a specialized audience and an interested public. So, let's see some more examples where the authors have successfully achieved this balance. We'll begin with a topic far-afield from climatology, namely cultural geography. In *GeoJournal*, Michael Broadway, Robert Legg, and Teresa Bertossi present findings on a study entitled "North American Independent Coffeehouse Culture: A Comparison of Seattle with Vancouver."[13] Unlike the specialized *Journal of Geophysical Research: Atmospheres, GeoJournal* is broader in scope. As stated on their home page, the journal is devoted to all branches of spatially integrated social sciences and humanities. This long-standing journal is committed to publishing cutting-edge, innovative, original, and timely research from around the world and across the whole spectrum of social sciences and humanities that have an explicit geographical/spatial component, in particular in *GeoJournal*'s six major areas: economic and development geography; social and political geography; cultural and historical geography; health and medical geography; environmental geography and sustainable development; and legal/ethical geography and policy.[14] Consequently, a broad readership that crosses many sub-disciplinary boundaries will necessitate that abstracts be written in such a way to convey the essence of the work in *relatively* non-specialized terminology.

Note that I underscore the *relativeness* of terminology. When negotiating the balance of specialized terms and concepts—the so-called charge of jargon—and plain language, we often have to make a judgment call. Many years ago, as a master's student in the geography department at San Diego State University, I was enrolled in a graduate seminar on research methods. We were required to write draft proposals and, in turn, critique our classmates' proposals. Often, discussion centered on when terms did or did not require clarification. I recall one seminar when the class evaluated a proposal submitted by a meteorology student. In this particular case, he used, but did not define, the term *insolation*. A number of cultural geographers thought that the term—insolation—should be defined. Perhaps not surprisingly, the physical geography students countered that the term was commonly understood and did not count as 'jargon'.

---

13 Michael J. Broadway, Robert Legg, and Teresa Bertossi, "North American Independent Coffeehouse Culture: A Comparison of Seattle with Vancouver," *GeoJournal* 85 (2020): 1645–1662.
14 *GeoJournal* home page, available at https://www.springer.com/journal/10708 (accessed April 13, 2021).

How did Broadway and his co-authors negotiate this balance in writing their abstract for *GeoJournal?* First, let's consider the title before reading the abstract. Upfront, we already learn that the study will be a comparison of coffeehouse culture in two locations: Seattle and Vancouver. These are two large urban areas, located in close proximity, but in separate, sovereign countries. We don't know *why* this particular study is important; nor do we know *how* the research was conducted. We also don't know if any particular theory or concept frames the study. Frankly, we may or may not fully understand what is meant by 'coffeehouse culture.' The title is informative but necessarily incomplete. Now, here's the abstract in its entirety:

Phenomenologists argue that place is central to human existence and that much of human behavior is habitual. A person who regularly visits a coffeehouse is likely to meet others engaged in a similar routine, over time they feel at ease and develop a sense of attachment to the space; a so-called third place—where people meet and engage in conversation. Since the early 1980s, the number of coffeehouses in North America has soared, but despite their ubiquity researchers have largely ignored whether they are third spaces. In the United States, critics charge that they have become places to be alone together, while in Canada, some research indicates that face-to-face conversation still flourishes in independent coffeehouses. This paper attempts to reconcile these competing perspectives by examining 30 coffeehouses in two neighborhoods in Seattle and Vancouver. Since design can affect social interaction, the coffeehouses are assessed on their spatial structure, how patrons use that space and how that space is assessed on social media. No difference was found between the coffeehouses in terms of their locational characteristics and how their physical environment was structured. However, a statistically significant difference was found in patron behavior. A majority of Vancouver customers engaged in face-to-face conservation, while Seattle patrons preferred to sit alone. Finally, Seattleite patrons were more likely to emphasize a coffeehouse's workplace function in their online reviews than their Vancouver counterparts. In sum, despite the two cities' close proximity there appears to be a difference between their residents in how they perceive coffeehouses, a workspace versus a social space.[15]

Broadway and his co-authors *suggest* that their research is framed in phenomenology. I qualify their statement, for they claim that phenomenologists make certain arguments about human existence but they don't explicitly say that they agree with this claim or that they frame their study accordingly. This is intimated and I certainly infer this; but at this point, it isn't determined. In addition, we can debate whether phenomenology itself should be defined. Is 'phenomenology' for cultural geographers akin to 'insolation' for physical geographers? I would probably say yes but, again, this is a judgment call. Notably, the authors quickly define their understanding of coffeehouse culture. They explain that since the 1980s, cof-

---

15 Broadway et al., "North American Independent Coffeehouse Culture," 1645.

feehouses have increased in number and are common places of human interaction. Next, the authors introduce—but do not expand upon—several key terms that we, as readers, would expect to be clarified in the main text, notably, place, space, third place. Broadway and his colleagues reference previous studies that indicate different forms of social behaviors in coffeehouses, depending on the geographic location. Procedurally, they utilize a mixed-method approach (implied) in a comparison of 30 coffeehouses in two neighborhoods. Statistical methods (unspecified) are used in the analysis; and findings indicate that patrons perceive the functions of coffeehouses differently. As a reader, what do I take away from this abstract? Does the abstract 'suffer' because it does not specify precise statistical methods or data collection techniques? Is the abstract weakened because we don't have a philosophical treatise on phenomenology? I would say no to both questions. Returning to Hart, abstracts are "concerned with ideas, not with facts and procedures."[16] In other words, unless the manuscript is explicitly designed to address a methodological technique, such as a particular statistical model or data collection instrument, the precise procedures are less relevant. Instead, the abstract underscores the main ideas under analysis, namely space and place. And on that point, the abstract sufficiently grounds the research for readers to determine whether the text is germane or not for a more in-depth reading. If one is currently interested in phenomenology, societal interactions at the micro level, or what makes a space become a place, the paper looks promising. My only quibble with this example? They write that their paper *attempts* to reconcile two competing perspectives.

My final example, an end-product abstract written by Stepha Velednitsky, Sara Hughes, and Rhys Machold, is for a paper entitled "Political Geographical Perspectives on Settler Colonialism" and appears in the journal *Geography Compass*.[17] *Geography Compass* is a unique journal, in that it provides summary reviews of research, aimed at both specialized and non-specialized audiences. As detailed on its home page, the journal publishes "original, peer-reviewed surveys of current research from across the entire discipline, with the aim of providing topical and significant research on a monthly basis. Geography Compass provides an ideal starting point for both specialist and non-specialist, offering pointers for researchers, teachers, and students alike, to help them find and interpret the best research

16 Hart, "Ruminations," 227.
17 Stepha Velednitsky, Sara N.S. Hughes, and Rhys Machold, "Political Geographical Perspectives on Settler Colonialism," *Geography Compass* 14, no. 6 (2020): e12490.

in the field."[18] Thus, the journal differs considerably in scope and purpose from either the *Journal of Geophysical Research* or *GeoJournal.*

From the title provided by Velednitsky et al., we have two important pieces of information. First, the manuscript is geared toward political geography and, second, the manuscript focuses on the concept of settler colonialism. Here's their abstract:

Given the centrality of land, territory, and sovereignty to settler colonial formations, it is unsurprising that geographers and other scholars working on such topics are increasingly finding settler colonial studies fruitful in their research agendas. However, work on settler polities in political geography has historically been marked by the present absence of this framework, which has been consequential in terms of circumscribing the kinds of political analysis that geographers can offer. It also limits the nature, depth, and scope of radical critique of violent domination by skirting certain questions about the core drivers of dispossession and responsibility for them. This article examines political geographical engagement (or lack thereof) across each of four themes: population management/governance, territory/sovereignty, consciousness, and narrative, paying particular attention to works which challenge the present absence of settler colonial theory in political geography. We argue that analyzing settler colonial formations as such is essential to conceptualizing their workings and linages or disjunctions with other forms of empire. Yet this focus also has broader political stakes related to geography's complicity with racialized state power, violence, and empire, as well as and [sic] efforts to decolonize the discipline.[19]

In terms of content, the abstract immediately raises several concepts—*land, territory,* and *sovereignty*; and, picking up from the title, *settler colonial formations* also makes an appearance. As someone who has read widely in political geography, I recognize that these are contested terms and have been subject to considerable debate. Would a meteorologist? Probably not. So, should the authors attempt, in the abstract, to provide further clarification of these terms? Probably not. As specified in the title, the manuscript clearly targets political geographers who, presumably, are well aware of the contested nature of these terms. Interestingly, a concept is introduced, that of a "present absence." In some sub-fields of geography, this is not an uncommon term and arguably can stand alone. However, the fact that it appears *twice* in the abstract suggests that it performs some important work in the manuscript. Moving on, the abstract asserts that geographers and other scholars interested in political geography, especially those working with the concepts of land, territory, and sovereignty, are engaging with settler colonial studies. Given the intended audience, settler colonial studies is not defined. However, the authors

18 *Geography Compass* home page, available at https://onlinelibrary.wiley.com/page/journal/17498198/homepage/productinformation.html (accessed April 13, 2021).

19 Velednitsky et al., "Political Geographical Perspectives," e12490.

do premise that the use of the term settler colonialism is uneven, in part because of the different ways it has been introduced and incorporated into geography. The lack of a precise definition, we infer, follows from the complexities of the term; hence a key contribution of the article. In addition, the unevenness of the term, its inclusion or exclusion, its presence or its absence, has, according to Velednitsky and her co-authors, limited the political analyses performed by geographers. In other words, the authors pose a critical conceptual problem surrounding the use of settler colonial formations. To that end, they review relevant scholarship in four broad contexts (themes) that engages with settler colonialism *to a greater or lesser degree.* Lastly, the authors indicate that how geographers approach the topic of settler colonialism has an importance that extends beyond any given research topic to include geography as a whole.

Collectively, the three end-product abstracts presented—Sheridan and Lee, Broadway et al., and Velednitsky et al.—illustrate key similarities, although the form and structure of the attendant articles are very different in scale and scope. Well-prepared abstracts enable readers to identify the basic content of a manuscript quickly and concisely, to determine its relevance to their interests, and thus to decide whether they need to read the manuscript in its entirety.[20] In all three examples, the authors effectively negotiate a balance between specialized terminology and general understanding and, in doing so, are able to clearly convey the essence of their research. And on this final point, it is helpful to think of abstracts not as vestigial appendages that merely appear before or after the primary manuscript. Instead, think of abstracts as their own, self-contained manuscripts in miniature. In other words, abstracts—similar to journal articles, book reviews, scholarly monographs, and the like—are meant to be read as coherent manuscripts in their own right. And if an abstract is to stand on its own, it is important the writer give the abstract appropriate footwear.

Some Final Thoughts on Writing Abstracts

As a work of scholarly activity, abstracts rarely receive the respect they deserve. And yet, of everything we write, abstracts are probably the most read form of academic scholarship. For that reason alone we should put more care into the drafting *and revising* of abstracts than we normally do. Too often we cut-and-paste disparate sentences (or sections!) from a finished manuscript; we cobble together

---

20 Ben H. Weil, "Standards for Writing Abstracts," *Journal of the American Society for Information Science* September–October (1970): 351–357; at 352.

some stray thoughts at the eleventh hour of manuscript submission. Intuitively, we probably know this to be inappropriate. In fact, I often draw parallels between writing abstracts and when I had to clean my room as a child. I didn't like cleaning my room and, having stored most of my clothes and toys in their proper place, I always seemed to have a little pile of 'stuff' that needed to go *someplace.* I think of abstracts like little piles of clutter that stand in the way of more important, more enjoyable activities, for until I cleaned it up, I couldn't go out to play. And so I crammed that pile under my desk, or under the bed, or in the closet: anywhere, just to get it out of sight from my parents. I *knew* that I wasn't really cleaning my room; and I know now that hastily pulling sentences together, almost at random, is not the most effective way to write something that will be seen by more people than the manuscript itself. And so, over the years, I try to do better, and rethink abstracts as tiny little papers—the ants of the academic world—that carry a terrific weight.

Chapter 3 Journal Articles

In geography, refereed journal articles have long been considered the coin of the realm. It isn't surprising, therefore, that many chapters and commentaries have appeared over the years, providing much-needed advice on preparing, writing, and submitting refereed journal articles.[1] Unfortunately, many of these well-intentioned pieces fall short of the mark. Too often, advice is offered on why we write articles, where and how often we should publish, how to co-author papers, how to submit manuscripts, and how to deal with rejections.

In a humorous piece that belies its seriousness, Bert Blocken identifies "10 Tips for Writing a Truly Terrible Journal Article."[2] These include:

1. Refuse to read the previous literature published in your field.
2. Take the lazy route and plagiarize.
3. Omit key article components.
4. Disrespect previous publications.
5. Overestimate your contribution.
6. Excel in ambiguity and inconsistency.
7. Apply incorrect referencing of statements.
8. Prefer subjective over objective statements.[3]
9. Give little care to grammar, spelling, figures and tables.
10. Ignore editor and reviewer comments.

These 'guidelines' are important, but they only skirt around the heart of the matter, that is, the format and structure of journal articles. Largely missing in 'how-to-write' chapters and commentaries are sustained engagements with the specific

---

1 Stanley D. Brunn, "Personal and General Publishing Policies of Geographers," *Terra* 99, no. 3 (1987): 155–165; Stanley D. Brunn, "The Manuscript Review Process and Advice to Prospective Authors," *The Professional Geographer* 40, no. 1 (1987): 8–14; Billie L. Turner, II, "Whether to Publish in Geography Journals," *The Professional Geographer* 40, no. 1 (1988): 15–18; David R. Butler, "Conducting Research and Writing an Article in Physical Geography," in *On Becoming a Professional Geographer*, edited by Martin S. Kenzer (Caldwell, NJ: The Blackburn Press, 2000), 88–99; and L.S. Bourne, "On Writing and Publishing in Human Geography: Some Personal Reflections," in *On Becoming a Professional Geographer*, edited by Martin S. Kenzer (Caldwell, NJ: The Blackburn Press, 2000), 100–112.

2 Bert Blocken, "10 Tips for Writing a Truly Terrible Journal Article," *Elsevier Publishing Campus*, March 1, 2017, available at http://www.urbanphysics.net/Elsevier_Publishing_Campus_Blocken_2017_PDF.pdf (accessed January 19, 2023).

3 This tip should be taken with a grain of salt. Blocken is a civil engineer and, while this recommendation applies to writing in scientific journals, it may make for a truly terrible article if you're writing for a humanities journal.

https://doi.org/10.1515/9783111189727-005

components, for example the sub-sections of journal articles, and how these sections hold together in a coherent manner. In this chapter I evaluate a selection of published scholarly articles to illustrate a range of writing techniques that can make for more effective manuscripts. Keep in mind, also, that many of the suggestions forwarded for the writing of journal articles apply to the writing of book chapters for edited volumes, scholarly monographs (books), book reviews, and encyclopedia entries. In the course of this and subsequent chapters, relevant differences are highlighted for additional discussion.

The Structure of a Journal Article

In high school I was required to write essays. Often, I was encouraged to structure my essay in a one–three–one format, that is, an introductory paragraph, three supporting paragraphs, and a concluding paragraph. The first paragraph centered the main topic, usually introduced with a 'hook' to get readers interested. I was told not to include too much information in the introduction. The 'meat' of the essay came in the three subsequent paragraphs. In these paragraphs, I was to provide three, independent, arguments in support of the main topic. Sometimes, we could add some variety and include arguments against the main topic. The conclusion provided a summary of the main topic and a review of the three source arguments.

When I entered California State University, Long Beach as an undergraduate student, I was required to write *term papers*. Essentially, these were expanded variations on my high-school essays. The differences were usually of scale and scope. In high school, completed essays were expected to be about three pages in length; in college, term papers were likely to run to about 15 pages. Occasionally—usually in history classes—I was required to produce a 30-page term paper. In terms of composition, my high-school essays had no headings and certainly no sub-headings. I was, though, required (sometimes for a grade) to produce an outline prior to writing my five-paragraph, three-page essay. At university, term papers did have headings—again, rarely sub-headings. Essentially, though, the main difference between my earlier essays and later term papers was one of length.

Graduate school changed all of this. At San Diego State, my professors demanded an altogether different format in my writing. I was no longer writing term papers but instead *seminar papers*. Here, the major change was that my 'thesis statement' be supported by the relevant literature. Admittedly, in high school I was required to include 'three to five' sources, properly formatted of course. And at university, also, term papers required 10–12 sources, also properly formatted. In fact, teachers frequently down-graded my work if I used an improper citation system or printed the paper in Calibri or Arial instead of Times New Roman. In my graduate program, these formatting principles remained, although professors were usually more lenient in grading. Instead, my professors in graduate school stressed the importance of properly situating my work within the large field.

I was encouraged to publish, certainly. But here's the problem—I was given little guidance in moving from writing seminar papers to journal articles.[4] As a graduate student, I continued to construct outlines prior to writing. I also made sure to identify the 'major topic' and to provide a 'guiding' thesis statement. My manuscripts were replete with references to (hopefully) the relevant literature. But when I compared my finished paper with the published articles I was required to read in seminars, I saw a terrific chasm. I was nowhere close to being able to write for an academic journal. What was I missing?

A lot, as it turns out.

Most salient for our immediate task, though, is valuing more fully the structure of research articles as opposed to essays, term papers, or seminar papers. This was a lesson I learned late in the process, as a doctoral student at the University of Southern California. I recall sitting in the computer lab one afternoon, struggling to improve my writing. That's not entirely correct. I felt comfortable with my writing *style*.[5] I was struggling with the organization of my writing. But on that day, Doug Sherman, a professor and coastal geomorphologist, told me to go through various journals and write out—for each article in each issue—the section headings, including any sub-headings. In hindsight, this exercise taught me to appreciate the different formats found in academic writing. I suppose at some level I understood this; by this point I had read countless journal articles. However, until Doug suggested that I deliberately outline each and every article, I remained hopelessly hamstrung by my earlier writing habits. From that day forward, I began drafting of all my manuscripts by first laying their structural foundation.

Framing the structure of a journal article can be one of the most important steps in the writing process. It can also be one of the most frustrating and anxiety-inducing steps. How to begin? What sections do we include or exclude? Normally, it's a safe bet to include an introduction and conclusion. But do I need a stand-alone 'literature review' section? What about a 'methods' section? Do I need a 'findings' section *and* a 'discussion' section?

The answers to these questions, as it turns out, depend almost entirely on the journal I want to target. Most academic journals, as microcosms of academic disciplines, operate within particular normative traditions. One simply does not write

---

4 Notice how I positioned my experience in this sentence within the broader literature!
5 As it happens, I really needed to improve my writing style also.

for *Nature* or *Science* in the same way that one might write for *Social Text*. This poses a problem for geographers, in that our 'discipline' straddles many different normative traditions. Consequently, before you begin drafting your manuscript, you should give some thought to possible outlets. Nothing is cast in stone and you can always revise a manuscript and submit to a different journal. That said, having a potential home for your manuscript before you begin can eliminate some of the anxiety in the writing process.

So how do you choose a journal? A glib response, not without merit, is to target those journals most appropriate to your work.[6] If you're researching and writing on economic geography, aim for *Economic Geography*; similarly, if you're doing work on political geography, target *Political Geography*.[7] Then again, you can target any number of 'generic' journals in geography, including the *Annals of the American Association of Geography, Transactions of the Institute of British Geographers, The Professional Geographer*, or *Area*. If you're working broadly as a human geographer, you might target *Progress in Human Geography*; if you're a physical geographer, you can look to *Progress in Physical Geography*; and if you're working on environmental geography, you might consider *Progress in Environmental Geography*. The possibilities to publish in geography are seemingly endless. Unlike decades past, there exists now a wide variety of journals, many of which target specific sub-fields. According to the website OOIR.org, there are approximately 85 'geography' journals in operation.[8] This is actually an undercount. OOIR.org is oriented toward the social sciences and does *not* include journals that publish work in physical geography or geospatial technologies. If these fields are included, the number of geography journals probably exceeds 200. And to these, we can add countless journals that publish in related fields.

So, how do you select a journal? This is, in fact, a complex process, a decision mediated by exogenous factors far beyond our control. Fundamentally, our writing practices, as Dydia DeLyser explains, "emerge from and are produced through our differing access to material and spatial resources—most foundationally (stable or precarious) academic employment and the income, office space, and time that may

---

6 Frank Witlox, "Getting Your Paper Reviewed and Finally Published in Journal of Transport Geography: The Do's and Don'ts from the Viewpoint of the Editor-in-Chief," *Journal of Transport Geography* 81 (2019), doi: 10.1016/j.trangeo.2019.102545.

7 Keep in mind that you probably won't be publishing in the same journal throughout your career. Let's say you've published a manuscript in *Political Geography*. Where do you publish your next?

8 "List of Journals—Geography," OOIR, available at https://ooir.org/journals.php?category=geography (accessed January 19, 2023).

(or may not) accompany our jobs (and the jobs we don't have)."[9] Natasha Webster and Martina Caretta agree, noting that we work in an uneven professional field that imparts a variety of experiences and means to both promote and foster early-career scholars.[10] When coupled with exposure to microaggressions for example that cut across intersectional subject-positions, including gender, class, sexuality, and ethnicity, writers are subject to intensified feelings of failure, isolation, exhaustion, anxiety, and stress.[11] When writing in this climate, something as banal as targeting a journal is fraught with challenges. The reality is, the choice made by a graduate student writing their first journal article differs greatly from that of a tenured professor writing their fiftieth journal article.

When I was a graduate student in the early 1990s, there was little conversation in geography on how to negotiate the neoliberal university. We were encouraged to publish in top-tiered journals. End of story. However well intentioned, we were instructed to know which journals were considered highly ranked and those considered secondary or tertiary. The overriding objective was to target the top journals and to build our reputations within and beyond our sub-fields.[12] And yet, as DeLyser observed long ago, most graduate students received little or no guidance in thinking through, organizing, and writing. Paradoxically, we were expected—required—to publish widely and often in leading journals but were *under-prepared* in cultivation of the skills and techniques necessary to publish.[13] Thankfully, there is now a growing body of work on how to conduct research and write within the neoliberal university and I encourage you to seek out these contributions.[14]

9 Dydia DeLyser, "Writing's Intimate Spatialities: Drawing Ourselves *to* our Writing in Self-Caring Practices of Love," *EPA: Economy and Space* 54, no. 2 (2022): 405–412; at 406.

10 Natasha A. Webster and Martina Angela Caretta, "Early-Career Women in Geography: Practical Pathways to Advancement in the Neoliberal University," *Geografiska Annaler: Series B, Human Geography* 101, no. 1 (2019): 1–6; at 2.

11 Puāwai Collective, "Assembling Disruptive Practice in the Neoliberal University: An Ethics of Care," *Geografiska Annaler: Series B, Human Geography* 101, no. 1 (2019): 33–43.

12 On this point, David Butler is particularly blunt. He wrote, more than two decades ago, "In a perfect world, you wouldn't have to worry about this, but you will because what your faculty colleagues think of your publication efforts will play a major role in how you are perceived in your department and how you are rewarded financially and by promotion." Butler, "Conducting Research," 91.

13 Dydia DeLyser, "Teaching Graduate Students to Write: A Seminar for Thesis and Dissertation Writers," *Journal of Geography in Higher Education* 27, no. 2 (2003): 169–181; at 169.

14 See for example Kye Askins, "Being Together: Everyday Geographies and the Quiet Politics of Belonging," *ACME: International E-Journal for Critical Geographies* 14, no. 2 (2015): 461–469; Lawrence D. Berg, Edward H. Huijbens, and Henrik Gutzon Larsen, "Producing Anxiety in the Neoliberal University," *The Canadian Geographer/Le Géographe canadien* 60, no. 2 (2016): 168–180; Kye Askins and Matej Blazek, "Feeling Our Way: Academia, Emotions and a Politics of Care," *Social*

But where does this leave us? In general, when preparing a manuscript and thus selecting a possible outlet, I ask myself: what do I want to say, and to whom do I want to say it? Let's say I'm developing a manuscript on Khmer Rouge irrigation projects built during the Cambodian genocide. This topic may be of interest to political geographers but also to geographers interested in water issues or infrastructure. In this case, a variety of journals might be appropriate. For example, I could target *Political Geography* or perhaps *Environment and Planning E: Nature and Space*. On the other hand, I could look beyond geography and publish in a more specialized journal such as *Genocide Studies International*.

However you ultimately decide on a potential journal, read through current and past volumes and get a sense of what types of manuscripts are published. Guidelines are readily available through the journal and/or its web-site, usually under the heading "Guidelines for Contributors." The *Annals of the American Association of Geographers*, for example, indicates that "manuscripts should be no longer than 11,000 words total, including abstract, references, notes, tables, and figure captions."[15] There is no specified format (i.e., a suggested template of sub-sections); in part, this lack of specificity is because the intended audience of the *Annals* is the totality of the geographic community. Flexibility is necessary to encompass the vast array of 'geographic' scholarship, and this speaks directly to writing style.

Let's delve a little deeper and contrast the *Annals* with *Geomorphology*. Here, the intended audience is more select: "*Geomorphology* publishes peer-reviewed works across the full spectrum of [geomorphology] from fundamental theory and science to applied research of relevance to sustainable management of the environment." Consequently, there is a greater specificity of the format of articles— although even here there is room for flexibility. According to the guidelines, "there are no strict formatting requirements but all manuscripts must contain the essential elements needed to convey your manuscript, for example Abstract, Keywords,

---

& *Cultural Geography* 18, no. 8 (2017): 1086–1105; Martina Angela Caretta, Danielle Drozdzewski, Johanna Carolina Jokinen, and Emily Falconer, "'Who Can Play This Game?' The Lived Experiences of Doctoral Candidates and Early Career Women in the Neoliberal University," *Journal of Geography in Higher Education* 42, no. 2 (2018): 261–275; Natasha Webster and Meighan Boyd, "Exploring the Importance of Inter-departmental Women's Friendship in Geography as Resistance in the Neoliberal Academy," *Geografiska Annaler: Series B, Human Geography* 101, no. 1 (2019): 44–55; and James D. Todd, "Experiencing and Embodying Anxiety in Spaces of Academia and Social Research," *Gender, Place & Culture* 28, no. 4 (2021): 475–496.
15 *Annals of the American Association of Geographers*, available at https://www.tandfonline.com/journals/raag21 (accessed April 16, 2021).

Introduction, Materials and Methods, Results, Conclusions, Artwork and Tables with Captions." Authors are expected to "divide the article into clearly defined sections."[16]

While not a template for the physical side of geography, the journal *Geomorphology* does reflect a fairly standard approach with conventional sub-headings. Articles appearing in *Antipode*, on the other hand, are often far afield from the 'scientific' model. As a self-titled journal of 'radical geography', articles in *Antipode* are frequently very theoretical—often (but not exclusively) informed by Marxism. It is understood, if not expected, that submissions to *Antipode* are, what I like to term, *free range*. By this I mean that authors have considerably more lee-way in the representation of their work. Indeed, the more idiosyncratic 'representation' (i. e., the writing) of the published work is often part-and-parcel of the 'analyses'.[17]

Having thus considered the outlet, let's look at the internal structure—the outline—of two very different manuscripts. The first is an article that appears in *Atmospheric Environment*,[18] and the second appears in *Antipode*.[19]

I. Introduction
II. Materials and Methods
   a. Data
   b. Methodology
III. Results
   a. ANOVA Results
   b. All-Pollutant AQI and Spike Days
   c. Primary Pollutant Analysis
IV. Discussion
   a. Contribution of WTs and CPs Separately to Total AQI
   b. Contribution of CP/WT Combinations to Total AQI
   c. Analysis of Individual Primary Pollutants
V. Conclusions

---

16 *Geomorphology*, available at https://www.sciencedirect.com/journal/geomorphology (accessed April 16, 2021).

17 Although this chapter is not the place to introduce the on-going philosophical debates in geography, it is worth noting that different theoretical approaches—such as critical race theory, queer theory, and feminist theory, to name but three—all entail different *relationships* between author, audience, and research, with notable differences in the process of writing and in the final representation of the manuscript.

18 Cameron C. Lee, Thomas J. Ballinger, and Natalia A. Domino, "Utilizing Map Pattern Classification and Surface Weather Typing to Relate Climate to the Air Quality Index in Cleveland, Ohio," *Atmospheric Environment* 63 (2012): 50–59.

19 Liz Bondi and Mona Domosh, "On the Contours of Public Space: A Tale of Three Women," *Antipode* 30, no. 3 (1998): 270–289.

I.   Introduction

II. "The Tunning of Elinour Rumming": Gender and Class in the Separation of Public and Private in Early Modern England

III. Sophie Hall Visits New York: Gender and Access to Public Space in the Mid-Nineteenth Century

IV. Towards Equality? Gender, Class, and Urban Space in the Late-Twentieth Century

V.   Conclusion

On the surface, the structure of these articles appears very different. The *Atmospheric Environment* article is presented in a fairly conventional 'scientific' format. And for those trained in the sciences—and especially for those who follow the 'scientific method'—readers will have no problem following the order of the manuscript. The introduction is composed of five paragraphs that situate the research in a larger context; a bifurcated section of 'data' and 'methods' follows. Results are presented without fanfare and, in the discussion section, interpreted more broadly. A brief conclusion wraps up the article, with mention made of both the limitations of the current research and opportunities for future work.

But what of the *Antipode* article? Clearly, there is no explicit 'methods' section; nor is there an immediately apparent 'analysis' or 'results' section. A closer inspection, however, reveals a simultaneity of 'method' and 'analysis'; and that for much *qualitative* research the representation (i.e., the writing) is a visual cue to the research process. What I mean by this can be most clearly understood conceptually. The classic 'scientific method', which is composed of hypotheses, experimental designs, data collection, data analyses, and findings, is very *linear.* Conversely, those approaches that are grounded in non-positivist philosophies are more *circular.*

Admittedly, a comparison between *Atmospheric Environment* and *Antipode* is a rather extreme example. However, the contrast is important because it reaffirms a fundamental aspect of geography, namely the importance of different scholarly conventions and traditions and—by extension—different writing styles. There is no one 'right way' to write scholarly articles. It is important, therefore, that prospective writers be familiar with these differences and to do their due diligence prior to writing for a particular journal.

Beyond more conventional research articles, geographers also write review essays. These, I should hasten to add, are different from book review essays, with the latter consisting of sustained reviews of two or more books. Here, comparisons are made between the books under review and, in turn, the books are placed in the wider context of the field. Review essays, conversely, are reflective pieces that provide an overview of a particular topic or sub-field. Within geography, a handful of journals are devoted to review essays, notably *Geography Compass, Progress in*

Human Geography, and *Progress in Physical Geography: Earth and Environment. Progress in Human Geography,* for example, "is the peer-review journal of choice for those wanting to know about the state of the art in all areas of human geography research—philosophical, theoretical, thematic, methodological or empirical. Concerned primarily with critical reviews of current research, *PiHG* enables a space for debate about questions, concepts and findings of formative influence in human geography."[20] *Progress in Physical Geography: Earth and Environment* serves a similar function, albeit for work on the physical side of the discipline. As indicated on its home page, for example, *PPG*

> is an international peer-reviewed journal, encompassing an interdisciplinary approach incorporating the latest developments and debates within Physical Geography and interrelated fields across the Earth, Biological and Ecological System Sciences. Contributions which review progress to date; which blend review material with new and original findings; or which introduce material, methods or techniques at the forefront of current knowledge, while setting directions for future work are welcomed. Authors need not be uncritically exhaustive in synthesizing research on a particular topic, but should concentrate on what they consider to be the most promising recent productive trends and developments which are likely to be transformative.[21]

Broader in scope, *Geography Compass*

> is an online-only journal publishing original, peer-reviewed surveys of current research primarily from the human side of the discipline. *Geography Compass* is inclusive: it does not privilege any one perspective over another, it is open to all authors, and publishes articles that are both theoretical and practical in orientation, or concerned with methodology, as well as issue-oriented reviews. The journal's emphasis is upon state-of-the-art reviews, supported by a comprehensive bibliography and accessible to an international readership of geographers and scholars in related disciplines. *Geography Compass* is aimed at students, researchers and non-specialist scholars, and will provide a unique reference tool for researching essays, preparing lectures, writing a research proposal, or just keeping up with new developments in a specific area of interest.[22]

In general, review essays serve a variety of purposes. In a helpful piece, Michael Bradshaw and Rochelle Lieber identify seven common objectives:
1. Recent debates.

---

20 *Progress in Human Geography,* "Journal Description," available at https://journals.sagepub.com/description/PHG (accessed April 16, 2021).
21 *Progress in Physical Geography,* "Aims and Scope," available at https://journals.sagepub.com/aims-scope/PPG (accessed April 16, 2021).
22 *Geography Compass,* "Aims and Scope," available at https://onlinelibrary.wiley.com/page/journal/17498198/homepage/productinformation.html (accessed April 16, 2021).

2. Areas where there has been a recent surge of interest, or substantial new developments.
3. Areas where developments in one corner of the field might speak to (or lead to) developments in another corner of the field.
4. Areas that have been neglected, but need to be revived (and the reasons for that).
5. Areas where there has been recent interest from the popular media and that might serve as the basis of debate in the classroom.
6. Comparisons of topics across different schools of thought.
7. Developments in other disciplines on a particular topic that is of interest to Geographers.[23]

Having established a general sense of review essays, let's compare three examples, one from each journal. First, a sample from *Geography Compass*. In his review essay on geography and capital punishment, Alex Colucci argues that "Killing and violence are critically important aspects of human life that can help us understand the condition of socio-spatial relations and politics across a variety of spaces and scales." However, Colucci identifies a gap in the literature: where, he asks, are our geographies of capital punishment? Asserting that "geographers have a great deal to offer," Colucci attempts to "intervene in existing debates [on capital punishment] and suggest some ways geographers can engage this phenomenon."[24] Colucci structures his manuscript around five main sections:
1. Introduction
2. Clarifying the Terms
3. Geographies of Capital Punishment
4. How to Approach the Geographies of Capital Punishment
5. Conclusions

Following a brief introduction that situates his review, Colucci in the second section clarifies and distinguishes among several key terms and concepts that are often (and erroneously) used interchangeably, for example judicial and extrajudicial executions, death penalty, and of course capital punishment. The third section is the longest and consists of several sub-sections and sub-sub-sections:
3. Geographies of Capital Punishment

---

23 Michael J. Bradshaw and Rochelle Lieber, "Review Essays," in *Publishing in Geography: A Guide for New Researchers*, edited by Alison Blunt and Catherine Souch (London: Royal Geographical Society, 2008), 14–16; at 15.
24 Alex R. Colucci, "Geographies of Capital Punishment: New Directions and Interventions," *Geography Compass* 14 (2020): e12548.

3.1   Mapping Capital Punishment
    3.1.1.   Space, place, location
    3.1.2.   Scale
    3.1.3.   The space–time politics of capital punishment
3.2   Bodies in Motion
    3.2.1.   Human and animal bodies
    3.2.2.   Mobilities
3.3   Politics of Access
    3.3.1.   Public and privatized executions
    3.3.2.   Peligion
3.4   Executing the 'Other'
    3.4.1.   Racism, racial capitalism, and settler colonialism

As a review essay, this is remarkably inclusive. Colucci addresses several general geographic terms, for example space, place, scale, and location, as these relate to capital punishment. In addition, Colucci highlights how capital punishment interrelates with several key concepts, including animal geographies, mobilities, racism, and religion.

Our second example is a 'progress report' published in *Progress in Human Geography.* Unlike review essays that appear, for example, in *Geography Compass*, progress reports are commissioned by the editors to provide critical summaries of research in the sub-disciplines of human geography. In practice, authors are commissioned to provide three progress reports, usually appearing over a three-year period. Here's the second progress report on "Geographies of Race and Ethnicity," written by Laura Pulido.[25]

1.   Introduction
2.   The Environmental Racism Gap
3.   Environmental Racism and Racial Capitalism
    a.   Producing difference and value
    b.   Operationalizing nonwhite devaluation
    c.   Environmental racism as state-sanctioned racial violence

Given that three progress reports are commissioned on a particular sub-field over a three-year period, authors have considerable flexibility in terms of representation. That is, some writers provide more conventional reviews of the literature, identifying key trends. Other writers address ongoing work but use the progress

---

25 Laura Pulido, "Geographies of Race and Ethnicity II: Environmental Racism, Racial Capitalism and State-Sanctioned Violence," *Progress in Human Geography* 41, no. 4 (2017): 524–533.

report to push the sub-field (or even geography as a whole) in a particular direction. Here, Pulido adopts the latter. She opens her report with the following assertion: "We need to rethink environmental racism."[26] Crisp and to the point. She elaborates that the environmental justice movement arose in the early 1980s and has been successful in key areas; however, environmental disparities between white and nonwhite communities, what she terms "the environmental racism gap," have not diminished and have probably worsened. On this point, Pulido uses her progress report "to reposition environmental racism so that it is recognized as fundamental to contemporary racial capitalism."[27] As such, Pulido not only provides a review of existing research, but she expressly adopts a normative position in her engagement with critical debates.

Our third and final example is a review essay on oil spill modeling, written by Jake Nelson and Tony Grubesic for *Progress in Physical Geography*.[28] Here, Nelson and Grubesic concede that there exists a number of literature reviews and syntheses that cover oil spill modeling, risk impact and vulnerability analysis, computational tools for evaluating spills, remote sensing techniques for monitoring and detecting spills, and oil spill response and real-time response techniques. However, Nelson and Grubesic adopt a different approach, that is, to "step back and holistically evaluate the evolution of oil spill modeling research as a knowledge domain."[29] In other words, Nelson and Grubesic want to review the literature more comprehensively than more conventional recitations of previous work, and opt instead to "shed light on the complexities and nuances of this research area, highlighting the many ways in which the sub-domains of oil spill modeling research are interconnected." In doing so, they "identify important scientific work and papers that help to span disciplinary boundaries and form the 'connective tissue' of this research domain" and thus "extrapolate and identify potential future trends."[30] Their review consists of five main sections:

1. Introduction
2. Overview of the Literature
3. CiteSpace
4. Results
5. Conclusion: The Path Forward

---

26 Pulido, "Geographies of Race," 524.
27 Pulido, "Geographies of Race," 524–.
28 Jake R. Nelson and Tony H. Grubesic, "Oil Spill Modeling: Mapping the Knowledge Domain," *Progress in Physical Geography* 44, no. 1 (2020): 120–136.
29 Nelson and Grubesic, "Oil Spill Modeling," 121.
30 Nelson and Grubesic, "Oil Spill Modeling," 121.

Unlike Colucci and Pulido's reviews, Nelson and Grubesic empirically evaluate the literature through the use of CiteSpace, a "bibliometric software package designed to promote a deeper understanding of knowledge domains."[31] Briefly, through a series of statistical techniques, including cluster analysis, Nelson and Grubesic 'model' the literature to demonstrate the overlap of scholarly work in various research domains. In other words, Nelson and Grubesic do not simply provide a literature review—they in fact provide a scientific analysis of the literature.

To conclude this section: there is no singular template for writing journal articles. Certainly, there are some components that are pretty common. Obviously (most) articles will have an introduction and conclusion; beyond that, there is tremendous diversity reflecting both disciplinary conventions and journal guidelines. The most helpful piece of advice: when writing for a particular journal, do your homework and read several issues not necessarily for content but instead for format. Indeed, take the time to outline the structure of individual articles, paying close attention to how these writers organized their material to present a coherent narrative that takes the reader from start to finish.

Writing an Introduction: The All-Important First Sentence

We all *know* what happens in the introductory section of an article: the author (or authors) introduces the subject of the article. What else needs be said? Actually, quite a bit more is necessary.

By way of example, I selected at random a copy of *The Professional Geographer.* The following first sentences all appeared in the May 2014 issue (volume 66, number 2):

> In the famous London Cholera Map of 1854, John Snow tested a waterborne germ theory by marking on a neighborhood map the locations of cholera illness incidents relative to various water pumps.[32]

> The market for biofuels is growing for a number of reasons, including rapidly rising fossil fuel prices, alternative fuel use targets, and national security issues.[33]

---

31 Nelson and Grubesic, "Oil Spill Modeling," 123.
32 E. Eric Boschmann and Emily Cubbon, "Sketch Maps and Qualitative GIS: Using Cartographies of Individual Spatial Narratives in Geographic Research," *The Professional Geographer* 66, no. 2 (2014): 236–248.
33 J. Christopher Brown, Lisa Rausch, and Verônica Gronau Luz, "Toward a Spatial Understanding of Staple Food and Nonstaple Food Production in Brazil," *The Professional Geographer* 66, no. 2 (2014): 249–259.

Following the Spanish overthrow of the Aztec capital Tenochtitlán in 1521, introduced livestock expanded throughout the Viceroyalty of New Spain, roughly equivalent to present-day Mexico.[34]

Job–housing relationships have received considerable attention in the urban geography literature.[35]

China's economic transition, starting in 1978 and now having a longer time span than its socialist era, has had a significant impact on its economic geography.[36]

There is a clear consensus among scientists, academics, and policymakers concerned with climate change that a radical reduction in carbon-based energy use—most markedly among the world's relatively wealthy, high-consuming populations—is necessary.[37]

Many projections forecast that biota will be exposed to temporally and spatially altered temperature and precipitation regimes and increased CO_2 concentrations in the coming decades.[38]

The year 2007 was among the driest on record in the Southeastern United States ..., and the U.S. Federal Emergency Management Agency (FEMA) estimated drought-related economic losses during that year to be well over $5 billion.[39]

Taken as a whole, these opening lines won't be confused with Kafka, Melville, or Morrison. But in terms of setting the stage, of introducing an important story, I think these are pretty good. Each, in its own way, identifies a problem. And whether we call this a plot, a hypothesis, or a thesis statement, the basic point remains the same: something is wrong and geographers are attempting to understand, explain, or resolve the problem. The first example hints at the problems of waterborne diseases; the second raises concerns over economics and national security; the third heralds the dangers of human-induced environmental change; and so on.

---

34 Richard Hunter, "Land Use Change in New Spain: A Three-Dimensional Historical GIS Analysis," *The Professional Geographer* 66, no. 2 (2014): 260–273.
35 Woo Jang and Xiaobai Yao, "Tracking Ethnically Dividing Commuting Patterns Over Time: A Case Study of Atlanta," *The Professional Geographer* 66, no. 2 (2014): 274–283.
36 Zhiqiang Liu, "Global and Local: Measuring Geographical Concentration of China's Manufacturing Industries," *The Professional Geographer* 66, no. 2 (2014): 284–297.
37 Joseph Nevins, "Academic Jet-Setting in a Time of Climate Destabilization: Ecological Privilege and Professional Geographic Travel," *The Professional Geographer* 66, no. 2 (2014): 298–310.
38 Ross J. Guida, Scott R. Abella, William J. Smith, Haroon Stephen, and Chris L. Roberts, "Climatic Change and Desert Vegetation Distribution: Assessing Thirty Years of Change in Southern Nevada's Mojave Desert," *The Professional Geographer* 66, no. 2 (2014): 311–322.
39 Jason T. Ortegren, Ashley Weatherall, and Justin T. Maxwell, "Subregionalization of Low-Frequency Summer Drought Variability in the Southeastern United States," *The Professional Geographer* 66, no. 2 (2014): 323–332.

Stylistically, these opening lines represent a variety of approaches; all of which can be effective. To understand why—and how—let's dig a little deeper into the *content* of introductory sections.

Writing the Introduction: To What Purpose?

Opening sentences can provide the springboard for an article.[40] This is a tall order for one sentence—no matter how impressive the sentence may be!—and thus requires substantial help. Other sentences are called on to assist in the writing of the introduction. This is not a free-for-all, however. These sentences are no less important but, and this is key, they must work together. As I repeatedly stress throughout, I strive for inner cohesion in my writing. I look for continuity, from sentence to sentence, paragraph to paragraph, and section to section. In traditional scientific writing, this cohesion and continuity is explicit; however, even in 'creatively constructed' manuscripts there remains a continuity and cohesion in presentation.[41] By way of example, let's start with Kevin Hetherington's (2004) article "Secondhandedness: Consumption, Disposal, and Absent Presence." Hetherington begins:

> The growing academic interest in consumption in recent years has, in large part, been a product of the cultural turn within social science. That is not to say that consumption as an issue was not there in the past. Thorstein Veblen's work on the new rich, or leisure class, in the United States at the end of the 19th century ..., Frankfurt School critiques of the culture industry ... and of the false needs created by capitalism ..., as well as writing on spectacle and the colonization of everyday life by commodification ..., have all made a significant contribution to our understanding of consumption and consumer culture. ... What has changed is that the more recent analysis of consumption has sought to valorize it as a creative source of cultural or in some cases individual expression, lifestyle, and taste rather than see it as a fetishistic expression of alienation and false needs.[42]

Hetherington opens with a declarative statement: "The growing academic interest in consumption in recent years has ... been a product of the cultural turn within social science." From the outset, Hetherington positions his article within a specific academic community. Moreover, he makes the statement—to be substantiated later—that there is in fact a growing interest in the subject-matter (i. e., consump-

---

40 Not all opening lines provide this function. However, I prefer to follow the old adage that you can only make a first impression once.

41 We could, however, discuss the 'postmodern' critique of continuity.

42 Kevin Hetherington, "Secondhandedness: Consumption, Disposal, and Absent Presence," *Environment and Planning D: Society and Space* 22 (2004): 157–173; at 157.

tion). In his second sentence, Hetherington takes a step back and lets us know that there is, however, a longer history of academic interest in the subject, and that his work should also be viewed as a continuation of this tradition. What I find really effective, however, is Hetherington's third sentence, where he writes: "What has changed is ...." In three sentences, Hetherington has foreshadowed a discussion of (1) contemporary debates; (2) how these debates fit into a larger research tradition; and (3) what is distinctive about his contribution. Take note that he has not (yet) revealed his specific thesis or contribution.

In the second paragraph Hetherington continues to frame his article. He begins with an acknowledgment of broader debates and, to an extent, competing factions within the field. Hetherington writes: "there have been serious differences of opinion within this field." The remainder of the second paragraph informs the readers of these debates while the third paragraph forwards a possible means of bridging the debate. According to Hetherington, "One way in which to understand these two recent approaches to consumption ... is to look at the point in the consumption process that they focus on." Here, Hetherington is letting the reader know that his paper should be viewed not as taking sides in the debate, but rather as forming a potential bridge between the bodies of literature. The fourth paragraph, consequently, provides background information on the advantages of such a position. It is not until the fifth paragraph that Hetherington indicates his particular argument. He begins by stating both what his paper is and is not. Hetherington writes: "My aim in this paper is not simply to suggest that the disposal part of the consumption process (as its end point) is something that has been neglected in recent work on consuming and to then redress that omission ... but instead to suggest that disposal be seen as a necessary issue integral to the whole process of viewing consuming as a social activity."[43]

Let's evaluate another example, this time from the sub-field of coastal geomorphology. Here, we have Robin Davidson-Arnott and Bernie Bauer introducing their research on aeolian sediment transport (wind-blown sand) on a beach.[44] They begin:

> Aeolian sediment transport across beaches is complex and characteristically variable in space and time. When wind speed is near the threshold of motion, the transport system is intermittent with long periods of inactivity punctuated by short bursts of sand movement. As wind increases, the transport intensifies in terms of the proportion of time that transport occurs and the mass of sand carried at any instant in time, but rarely does it reach steady state equi-

---

43 Hetherington, "Secondhandedness," 158.
44 Robin G.D. Davidson-Arnott and Bernie O. Bauer, "Aeolian Sediment Transport on a Beach: Thresholds, Intermittency, and High Frequency Variability," *Geomorphology* 105 (2009): 117–126.

librium. When the mean wind speed is well above the threshold, transport becomes continuous, but it typically continues to exhibit considerable unsteadiness in terms of mass flux. ... The presence of surface moisture on all or part of the beach often contributes to enhanced intermittency or unsteadiness because stripping of thin layers of dry sand (conditioned by rapid evaporation) from the surface exposed moist, cohesive sand layers below, leading to surface patchiness. Although these general controls are understood conceptually, relatively little quantification of the character of intermittency has been conducted.[45]

Davidson-Arnott and Bauer's introductory paragraph is, on the surface, exceptionally technical and, at first blush, appears quite different from Hetherington's opening. So what, exactly, are Davidson-Arnott and Bauer *doing*? They begin with a seemingly banal statement that the movement of wind-blown sand across beaches is complex and differs over space and time. They proceed to describe, conceptually, the factors that influence sediment transport. They conclude this opening paragraph with a declarative statement that announces clearly their task at hand, specifically that "relatively little quantification of the character of intermittency has been conducted."[46] I really like this opening, for a couple of reasons. On the one hand, I'm struck by the straightforward presentation of a complex subject. Not being trained in aeolian geomorphology, I can't pretend to fully understand some key terms, for example the "threshold of motion" and the "unsteadiness in terms of mass flux." These terms, however, should not be considered jargon and my ignorance doesn't require that Davidson-Arnott and Bauer define these terms. They are writing for a specialized journal, one whose readership can be expected to be quite comfortable with these concepts. That said, even I as an 'outside' reader can follow their logic: conceptually, this is what happens with wind-swept sand; several factors influence this movement, including soil moisture and wind speed. We *know* these factors are important, but it is really difficult to measure these factors. Thus, having announced the problem, Davidson-Arnott and Bauer proceed to tell their story, that is, "to examine thresholds of sand movement, intermittency, and the relationship between fluctuating winds and transport intensity on the beach."[47]

---

45 Davidson-Arnott and Bauer, "Aeolian Sediment," 117–118.
46 Davidson-Arnott and Bauer, "Aeolian Sediment," 118.
47 Davidson-Arnott and Bauer, "Aeolian Sediment," 118.

Introductions and Writing to an Audience

It is important to select appropriate journals and to target specific audiences prior to writing. Frank Witlox, for example, recommends that you should "carefully select the most appropriate journal" and, having made that decision, "write your paper based on the journal guidelines."[48] This is sound advice; however, I would like to expand on this recommendation. Simply put, any given topic may be appropriate for a variety of journals and audiences. I can't tell you which to choose but I can provide some insight into how these impact how you write an introduction.

By way of illustration, much of my earlier work addressed the gendering of Philippine international labor migration. Simply stated, this subject appears somewhat narrow: of importance to scholars working, obviously, on gender and Philippine international labor migration. However, when writing manuscripts, I was actually in conversation with at least three (sometimes distinct) audiences and this affected how I wrote. Specifically, I wanted to write to those scholars interested in migration broadly and, specifically, international migration; to those scholars interested in the Philippines (but especially on migration-related issues); and to those scholars interested in feminist approaches to migration. Consequently, *how* I introduced my article was determined in part by the particular audience (and journal) to which I was writing. For a journal whose readership was regional, for example the *Philippine Population Review* or *Asian and Pacific Migration Journal*, I would introduce the article with specific reference to the Philippines and connect this to the broader topic of migration. In my *Philippine Population Review* article on scales of sexuality and the migration of Filipina overseas contract workers, for example, I write:

> The magnitude of labor export from the Philippines, both in volume and geographic scope, is without parallel. In 2000 alone, a total of 841,628 migrant workers from the Philippines were legally deployed; spatially, they found employment in over 160 countries and territories. Apart from the sheer size of the Philippines' overseas employment program, an additional noticeable feature is the predominance of female migrants. In 2000, for example, nearly 72 percent of all newly-hired contract workers from the Philippines were women. Patterns of Philippine overseas employment, thus, are consonant with other identified systems of international migration, namely, the increased 'feminization' of transnational mobility.[49]

My objective in this opening paragraph is to focus explicitly on migration patterns in the Philippines and, secondarily, to cast the patterns of Philippine migration

---

48 Witlox, "Getting Your Paper Reviewed."
49 James A. Tyner, "Scaled Sexuality and the Migration of Filipina Overseas Contract Workers," *Philippine Population Review* 1, no. 1 (2002): 103–123; at 103.

within a larger, international context. I do not address any academic debates in this paragraph; nor do I refer to the voluminous literature that exists on the subject. In the second paragraph, however, I do step back to draw on broader literatures in my subsequent discussion of Philippine migration. Hence, the second paragraph begins:

> As one of the most striking economic and social phenomena of recent times, the feminization of international labor migration raises crucial policy issues and concerns. ... [The] status of female migrant workers—as women, as migrants/non-nationals, and as workers in gendered segregated labor markets—makes them particularly vulnerable to various forms of discrimination, exploitation, and abuse. Accordingly, scholars of migration have acknowledged more so than ever the political ramifications of these migration trends.[50]

Here, I connect back to my initial opening of the article, that is, I argue that gendered patterns of Philippine international labor migration are part of a larger 'feminization' of mobility and that women's position in the global labor market places them in a particularly vulnerable position. Thus, the 'scale' of the paper moves from the specific (the Philippines) to the general (gendered labor migration).

In other articles, I reversed the organization, leading with the general and then focusing on the specific. In "The Gendering of Philippine International Labor Migration," published in *The Professional Geographer*, I wanted to speak to the study of gender and migration more broadly while using the Philippines as a case study. Since I was targeting *The Professional Geographer*—a journal with a wide readership—I framed my introduction to appeal to a wider audience than I did for the *Philippine Population Review.* My opening paragraph reads:

> The study of women in migration has paralleled advances made within the larger field of women's studies, shifting from an emphasis on the identification of sex-based patterns of migration to the explanation of those patterns. During the 1970s and 1980s, researchers sought to document the existence of women in migration flows ..., a task that is ongoing. This work uncovered strong sex differences in migration flows and challenged male biases in research, including the assumptions that patterns of female migration mirror those of male migration, and that men are the initiators of migration where women are merely followers.[51]

I continue along this path in the opening paragraph, identifying relevant literatures and current debates on the scholarly study of women in migration. Nowhere

50 Tyner, "Scaled Sexuality," 103.
51 James A. Tyner, "The Gendering of Philippine International Labor Migration," *The Professional Geographer* 48, no. 5 (1996): 405–416; at 405.

in the initial paragraph do I mention 'the Philippines'. Indeed, it isn't until the third paragraph that I identify the geographic focus of the article. I write: "I provide an institutional analysis of the Philippine labor migration industry to explain the production of gendered patterns of international labor migration. Specifically, I focus on how peoples, places, and occupations are represented by government and private institutions in the labor migration process."[52] Only later do I explain why I'm using the Philippines as a case study: "The Philippines provides a remarkable opportunity to examine the gendering of formal contract labor migration. It is the largest Asian exporter of labor, and exhibits a strong sexual division of labor flows."[53]

In both articles, I cover similar terrain; and to a degree, similar information is provided. However, the presentation of material differs starkly. Here's the takeaway. When targeting specific audiences, remember to structure your introductory material in an appropriate manner. This often requires a sensitivity to the scale of the particular manuscript you're writing. In some instances, it may be appropriate to present material from a more universal scale and work toward the particular, and in other situations, it's better to work from the ground up, so to speak. Essentially, I try to keep sight of the forest from the trees, that is, to remain sensitive both to the specifics (the trees) and the larger question (the forest). Sometimes we begin by looking at the forest and focus on individual trees; other times we identify trees and step back to view the forest in its entirety.

Writing the Introduction: Setting the Tone

Apart from presenting the subject-matter, introductory sections are important in establishing a particular tone—or feeling—for the article. This, of course, differs based on writing format. Those writers who adopt, by choice or necessity, a more technical format generally represent their material in a straightforward manner. This does not mean, however, that writing must be dry and boring; nor does this imply that writing must be robotic. A good example of *engaged* scientific writing appears in an article on subglacial hydrology, published by Mandy Munro-Stasiuk, Timothy Fisher, and Christopher Nitzsche in *Quaternary Science Reviews.*[54]

---

52 Tyner, "The Gendering of Philippine," 405–406.
53 Tyner, "The Gendering of Philippine," 406.
54 Mandy J. Munro-Stasiuk, Timothy G. Fisher, and Christopher R. Nitzsche, "The Origin of the Western Lake Erie Grooves, Ohio: Implications for Reconstructing the Subglacial Hydrology of the Great Lakes Sector of the Laurentide Ice Sheet," *Quaternary Science Reviews* 24 (2005): 2392–2409.

Here, the authors frame their specific case study within the wider debates of the sub-field:

> Glacial grooves have been identified on a number of paleo and modern glacial surfaces world-wide. ... The term 'groove' loosely describes narrow linear troughs ranging from metres to tens of kilometres long, with forms longer than approximately one hundred meters frequently called mega-grooves ..., or megalineations. At all scales, groove formation has been attributed to abrasion and excavation by clasts in the underside of glaciers and ice sheets ..., by subglacial squeezing and deformation of till ..., by subglacial slurry erosion ... and by subglacial meltwater erosion. ... We believe that insight on the precise mechanism of groove formation will ultimately help us better understand, and reconstruct, subglacial processes during the time of groove formation.[55]

In this paragraph all of the expected bits-and-pieces of an introduction are present. Munro-Stasiuk and her co-authors begin by noting the widespread existence of grooves; provide a working definition of this crucial term; and identify the ongoing debate as to groove formation. They identify also why their study is important.

On the 'human' side of geography, writing styles are somewhat more diverse. Academic articles can, of course, be straight-to-the point, similar for example to the openings provided by Davidson-Arnott and Bauer and Munro-Stasiuk and her co-authors. Here's the introductory paragraphs of Wei Li, Canfei He, and Huaxiong Jiang's article "Spatial and Sectoral Patterns of Firm Entry in China," appearing in *The Professional Geographer.*[56] They write: "Economic development is a dynamic process with firm entry, growth, and exit. New firms usually introduce new technologies, creatively recombine resources, and exert pressure on incumbent firms to improve their productivity, thus serving as a key driver for economic growth and employment creation." In turn, following Li and her co-authors, there has been extensive investigation into the patterns and determinants of firm entry in economic geography and regional economics, with two main strands of research apparent. Stylistically, the authors describe a real-world process and highlight that interpretations of this process have bifurcated into two dominant explanations. Negotiating these two approaches, Li et al. "investigate how the spatial pattern of new firm entry is related to industry structure change and firm heterogeneity."[57] Clear and concise,

---

55 Munro-Stasiuk et al., "The Origin of the Western Lake Erie Grooves," 2392.
56 Wei Li, Canfei He, and Huaxiong Jiang, "Spatial and Sectoral Patterns of Firm Entry in China," *The Professional Geographer* 71, no. 4 (2019): 703–714.
57 Li et al., "Spatial and Sectoral Patterns," 703–704.

this opening engages immediately both with the real-world issue and the geographic study of the problem.

Other introductions are more affective, that is, framed to draw readers emotionally into a problem. There are risks involved in this approach. There is a danger, for example, of appearing to be voyeuristic, that is, exploiting the suffering of others to make a point. When done effectively, however, it is possible to generate empathy in the reader; to help the reader understand that our research is *not* detached but does immediately affect—and is affected by—real people with real problems. Writing in *Environment and Planning D: Society and Space*, Adam Bledsoe and Willie Wright introduce a sense of urgency in their paper on "The Anti-Blackness of Global Capital."[58] They begin:

> The world is living through a moment in which hyper-visualized examples of anti-Black violence have gripped the public and spurred discourses around Black Life and its prevailing (lack of) value. ... From Ferguson to Baltimore to Charleston to New York to Minneapolis to Orlando to Rio de Janeiro, highly publicized images of the murder of Black women, men, children, and transgendered people have forced academics and lay people alike to reflect on the material and immaterial factors that have created the world in which we currently live.[59]

This is a powerful opening, one that immediately draws the reader into a world rife with anti-Black violence. The *geography* of this violence is palpable—and not simply because it occurs in the locations identified; rather, as Bledsoe and Wright intimate, violence is pervasive in the varied spaces and places of everyday life. Madeleine Hamlin adopts a similar approach and establishes a comparable tone in her work on public housing and prisoner reentry in Chicago.[60] However, unlike Bledsoe and Wrights' more expansive opening of a planetary violence, Hamlin personifies a form of structural violence. She begins:

> By July of 2016, Jimmy Beaman, Bobby Flowers, and John Stamps had all been on the waitlist for Chicago public housing for several years. All three have criminal records and had been living in a transitional housing facility on Chicago's west side since their respective releases from the Illinois Department of Corrections. These three men were among the first group of formerly incarcerated individuals to be accepted into Chicago Housing Authority (CHA) housing as part of a pilot program launched in November 2014 that aimed to overturn a ban on

---

58 Adam Bledsoe and Willie Jamaal Wright, "The Anti-Blackness of Global Capital," *Environment and Planning D: Society and Space* 37, no. 1 (2019): 8–26.

59 Bledsoe and Wright, "The Anti-Blackness," 8–9.

60 Madeleine Hamlin, "Second Chances in the Second City: Public Housing and Prisoner Reentry in Chicago," *Environment and Planning D: Society and Space* 38, no. 4 (2020): 587–606.

anyone with a criminal record from living in public housing. And yet, nearly two years later, they waited.[61]

In this opening paragraph, Hamlin directs her attention toward three real-life people and *shows* the problem without explicitly naming the problem. Effectively, Hamlin draws on the life-experiences of Beaman, Flowers, and Stamps to humanize the research problem. However, to clarify the issue at hand in no uncertain terms, Hamlin follows this introductory paragraph with the statement, "In this paper, I use this CHA pilot program as a case study and entry point to examine the complex entanglements of the housing and prison systems at the scale of the city—entanglements that speak to broader political and economic structures that transcend Chicago, though Chicago's experience may not be directly transferable to other contexts."[62] For Hamlin, this second sentence performs a heavy task. First, Hamlin articulates the main focus of her manuscript, namely how people become entangled in housing and prison systems. Second, she identifies the scale of analysis, that is, the city level. Third, she steps back, suggesting that the example of Chicago is not necessarily unique; indeed, the ways in which housing and prison systems interact with political and economic structures in Chicago may be found elsewhere. That said, Hamlin remains sensitive that places are unique and that caution must be taken when making geographic comparisons. All the while, we as readers remain mindful that the lives of real people are at stake.

Early in my career, I used an approach similar to that of Bledsoe, Wright, and Hamlin to humanize an otherwise staid discussion. Working on the bureaucracies of overseas employment agencies in the Philippines, my immediate focus centered on the unintended violence that is manifest in labor migration policies. Indeed, the bulk of the analysis was a discourse analysis of employment contracts, procedural reports, and other government documents. However, I was acutely aware of the human dimensions of overseas employment and, similar to Hamlin's portrayal of Beaman, Flowers, and Stamps, I felt it necessary to bring an empathetic reading to an otherwise faceless analysis. I opened the manuscript with the death of Maricris Sioson, a young woman migrant who died in Japan:

Maricris Sioson, a Filipina, was deployed to Japan as a performing artist on April 27, 1991. She left with a valid 6-month 'entertainer' visa, and worked as a dancer in a nightclub in Fukushima. Beginning in late August, Sioson complained of chronic fatigue and was anorectic. She was admitted to Hanawa Welfare Hospital on September 7 and died 7 days later. The imme-

---

61 Hamlin, "Second Chances," 587–588.
62 Hamlin, "Second Chances," 588.

diate cause of death was officially listed as multiple organ failure arising from fulminant hepatitis. Maricris was 22 years old.[63]

As an opening paragraph, there appears to be little academic or scholarly information; I don't refer to any intellectual debate, nor do I provide a thesis statement. A closer reading, however, presages my argument, for in my narrative of Maricris, I relate her migration and subsequent death to myriad administrative and institutional factors. Notice, for example, that I do not say she *migrated* or *moved* to Japan, but instead was *deployed*. As I explain later in the manuscript, contract labor migration is a big business and men and women are, to put it bluntly, treated as commodities. They are also, however, strategic 'resources' for labor-exporting governments and thus are *deployed* as one might dispatch a reserve army of surplus labor. Also apparent in the first paragraph are references to visas and an 'official' cause of death—itself foreshadowing that perhaps her death was not all that it seemed. The second paragraph picks up this theme:

> Doubts quickly surfaced in the Philippines about the cause of her death. The remains of Sioson revealed numerous cuts and bruises, and at the request of family members an autopsy was performed by medical personnel in the Philippines. While results did indicate the beginning stages of hepatitis, the cause of her death was attributed to traumatic head injuries.[64]

The controversy surrounding Sioson's death is continued in the third paragraph. It is not until the sixth paragraph that I state my main thesis:

> I present a case-study of the events following the death of Maricris Sioson and utilize this incident to document how specific notions of gender and sexuality are incorporated into the construction and reconstruction of policy. In particular, I examine how female migrants are, in the abstract, reclassified based on socially constructed notions of sexuality and, subsequently, morality/immorality. ... In the process, I demonstrate how the representation of exploitation within systems of labor migration serves the purposes of dominant factions of society, with little regard to the actual lived experiences of migrant workers.[65]

Tone is important. For Bledsoe, Wright, Hamlin, and myself, we deemed it necessary to immediately challenge the bureaucratic tendency to de-personalize and dehumanize violence at whatever scale of analysis. In doing so, we established a more empathetic and humanized tone to our work. Each of these articles, I suggest,

---

63 James A. Tyner, "Constructing Images, Constructing Policy: The Case of Filipina Migrant Performing Artists," *Gender, Place and Culture* 4, no. 1 (1997): 19–35; at 19.
64 Tyner, "Constructing Images," 19.
65 Tyner, "Constructing Images," 20.

would be less effective if they had been written in a detached, technical format. Accordingly, as writers we should think deeply and reflect seriously on the tone we want to establish. From the opening sentence onward, you only have one chance to make a first impression. How the introductory section is presented will greatly condition how readers approach the remainder of the manuscript.

Methods, or 'The Devil is in the Detail'

> No methods are discussed, but I didn't miss them. This appears as a literary and historical reflection, along with a research visit to the museum, on which the author describes its displays. Given the content and discussion, I thought this was ok. The article is written in a literary style, and adding a section on methods, unless intertwined in the text, would seem a jarring intermission that doesn't fit with the style.[66]

These comments were provided by Reviewer #2 on a recent manuscript I submitted for publication.[67] The content of the manuscript, at this point, isn't important. What is important is that the reviewer recognized that the style of writing was intimately connected to the contents of the manuscript, notably the inclusion or exclusion of a methods section.

Not all journal articles require methods sections. Indeed, many geographers now advocate for writing styles beyond the traditional, predominantly technical, formats. And, in many instances—including work in the physical sciences—more creative approaches are not only warranted but also welcomed. That said, the need to write in more convention formats remains, and this requires the inclusion of stand-alone sections on methods.

In general, research methodologies comprise two key elements: data collection procedures and data analysis. Methods sections in journal articles should, reasonably, provide clarity for both components. Let's start with Davidson-Arnott and Bauer's study of aeolian sediment transport. They write:

> The data described in this paper are derived primarily from measurements made across one beach profile established perpendicular to the shoreline. Wind speed measurements were made using RM Young cup anemometers with continuous DC output. A vertical array of cup anemometers was established on a tower in the mid-beach area, with anemometers

---

66 Review comments from Reviewer #2, *Austrian Journal of Historical Studies*, personal correspondence.

67 Despite the widespread belief among academics, Reviewer #2 is not always grumpy and snarky.

mounted at heights of 0.25, 0.6, 1.1, 1.9, and 2.6 m. Wind speed and direction were measured at the top of the mast at a height of 4.15 m using a Gill Windsonic anemometer.[68]

Wei Li and her co-authors include a similar section in their empirical paper on spatial and sectoral patterns of firm entry in China. They write:

> The data used in this article are from the *Annual Survey of Industrial Firms of China* (1998–2008), which was collected and administered by the National Bureau of Statistics of China. The database covers all Chinese firms including state-owned enterprises (SOEs) and non-SOEs with annual sales above 5 million RMB. Each firm item contains information such as firm identification, address, inventory, output, employment, register year, record year, ownership, and so on. We mainly selected manufacturing industries with four-digit standard industrial classification codes from 1300 to 4190 and 414 four-digit industrial sectors was retained as our analytic data set.[69]

Li and her co-authors continue, specifying in considerable detail how the selected data sets were classified and which statistical techniques were employed (i. e., Gini coefficient and Moran's *I*).

In both of these examples, data collection procedures and data analyses are presented clearly and concisely. This is a hallmark of much empirical work and of quantitative methods in general. Indeed, this information is required so that future research can replicate or validate the findings. But what of more qualitative studies, such as Hamlin's work on public housing and prisoner reentry? A notable difference—at least, in this example—is that the 'methods' section (1) appears much later in the manuscript and (2) is not set apart as a stand-alone 'methods' section. For Davidson-Arnott and Bauer, the organizational structure begins: Introduction, Study Area, Methodology.... For Li et al., the structure is similar: Introduction, Literature Review, Data Sources and Methods .... Hamlin, however, adopts a different structure: Introduction, The Carceral City, Public Housing as Prison?, From Locked In to Locked Out, Finding a Home in the Carceral City, Policy Logics, Conclusion. Far from a conventional, technical format, Hamlin is expressly writing a narrative, that is, her writing carries the reader along from concept to concept. This differs from the 'scientific' writing structure (and styles) of the two previous examples. In both of those studies, of aeolian sediment transport and the Chinese firms, the structure of the manuscript replicates the scientific method; this requires stand-alone sections on data sources and methods (and, of course, findings, and so on). With Hamlin's manuscript, however, the narrative takes center-stage, as opposed to the research process.

---

68 Davidson-Arnott and Bauer, "Aeolian Sediment," 118.
69 Li et al., "Spatial and Sectoral Patterns," 705.

And yet: *there is a method underlying Hamlin's research!* Let me make this perfectly clear: the lack of a stand-alone methods section does not mean that no methodological procedures were performed by the researcher. Instead, it illustrates a variation in presentation, one that is sensitive to the different epistemological assumptions inherent to 'qualitative' or 'quantitative' research. This isn't the place to engage in these philosophical discussions; there are several outstanding reference sources available, including sources specific to geography.[70] My point, rather, is to underscore that the structure of one's manuscript should be consonant with the philosophical grounding of the research process. This is an important lesson, one that is regrettably too often neglected, as Katherine Burlingame makes clear.[71]

Returning to Hamlin's study, how does she incorporate her methodological discussion? Deep into the manuscript, Hamlin eventually explains that she conducted "in-depth, semi-structured interviews with 13 individuals involved in the [CHA] housing pilot program" in order to "learn about stakeholders' goals for the project as well as what they saw as possible limitations."[72] She elaborates that key documents were also collected, including application forms and other materials required of potential participants. In this way, Hamlin provides sufficient detail without detracting from her narrative.

Qualitative studies *can* be more detailed in their specification of research methods. Considerable detail can be included with regards to interview or ethnographic techniques, archival methods, and analytical techniques (e.g., the various forms of discourse analysis). When in doubt when writing one's manuscript, read other articles. If you're writing results based on a year-long ethnography and are unsure how you can represent your research *without* detracting from your narrative, look at other, comparable studies. Just remember, there is no grand template to structure your manuscript.

---

70 See for example John W. Creswell, *Qualitative Inquiry and Research Design: Choosing Among Five Traditions* (Thousand Oaks, CA: SAGE Publications, 1998); Norman K. Denzin and Yvonna S. Lincoln (eds), *Collective and Interpreting Qualitative Materials* (Thousand Oaks, CA: SAGE Publications, 1998); and Norman K. Denzin and Yvonna S. Lincoln (eds), *The Landscape of Qualitative Research* (Thousand Oaks, CA: SAGE Publications, 1998). For discussions specific to geography, see Dydia DeLyser, Steve Herbert, Stuart Aitken, Mike Crang, and Linda McDowell (eds), *The SAGE Handbook of Qualitative Geography* (Thousand Oaks, CA: SAGE Publications, 2009).
71 Katherine Burlingame, "Where are the Storytellers? A Quest to (Re)enchant Geography Through Writing as Method," *Journal of Geography in Higher Education* 43, no. 1 (2019): 56–70.
72 Hamlin, "Second Chances," 593.

Placing Geographic Research and Writing: The Study Site

In geography, place matters. But does place matter in journal articles? Asked differently, do we need to include a geographic-based section on the study site? The predictably simple and not immediately helpful answer is, yes and no.

To clarify my non-answer, I'll start with an example from physical geography. Returning to

Davidson-Arnott and Bauer's work, we find after their detailed introduction a stand-alone section entitled simply 'Study Area' and a corresponding map. They write:

> The study was carried out at Greenwich Dunes, Prince Edward Island National Park, Canada (Fig. 1). Greenwich Dunes is located to the east of St. Peter's Bay and consists of about 4 km of shallow sandy beach and dunes overlying relatively weak sandstone bedrock (Fig. 1). The study site was located about 1 km east of the estuary mouth where the beach is 30–40 m wide and the foredune is about 8 m high with a relatively straight, simple stoss slope and more complex crest. ... Beach sediments are dominantly quartz sand with a mean diameter of 0.26 mm. The area is micro-tidal with a mixed semi-diurnal regime and a maximum range at spring tides of about 1 m.[73]

Davidson-Arnott and Bauer continue on, but you get the point. Detail is important, and the greater the specificity the better. Throughout physical geography, 'study site' sections are common. These don't have to be long but, in many cases, their inclusion is vital. What about human geography?

There are beaches and sandy dunes in Cambodia. Beyond that, I can't say much about the composition or size of the beach sediments. And for most of my work, this level of detail is not important. Geography is important, though, just at a different scale. In our work on irrigation systems constructed under forced labor during the Cambodian genocide, for example, my co-authors and I find it necessary to provide in detail the hydrology and geomorphology of Cambodia. Perhaps not to the level required of a geomorphologist (we don't, for example, specify the mean flow of the Tonle Sap River), but we do highlight the influence of topography and climatology and why this is important in understanding the construction of dams, dikes, canals, and reservoirs under the Khmer Rouge.[74] Structurally, we don't always include a stand-alone section on

---

73 Davidson-Arnott and Bauer, "Aeolian Sediment," 118.
74 See for example James A. Tyner, Mandy Munro-Stasiuk, Corrine Coakley, Sokvisal Kimsroy, and Stian Rice, "Khmer Rouge Irrigation Schemes during the Cambodian Genocide," *Genocide Studies International* 12, no. 1 (2018): 103–119; Corrine Coakley, Mandy Munro-Stasiuk, James A. Tyner, Sokvisal Kimsroy, Chhunly Chhay, and Stian Rice, "Extracting Khmer Rouge Irrigation Networks from

the 'study site'; indeed, the 'study site' is effectively the study. My point is, there are different ways that geographers can (and should) position their writings as appropriate.

In our writing, we must decide what is required to communicate our argument *to the specific audience targeted.* In our work on Khmer Rouge irrigation schemes, for example, we find ourselves alternating between speaking to genocide scholars and experts in remote sensing. Consequently, it matters how much geographic detail we provide up front. When publishing in *Remote Sensing*, for example, we lean toward a greater level of detail and may provide a stand-alone section on Cambodia's hydrology and geomorphology. Conversely, this level of detail on both the methods and study site is condensed considerably when we target journals such as *Genocide Studies International.* I should note, also, that sustained discussions of the genocide are minimized in our work appearing in *Remote Sensing* compared, obviously, with our manuscripts published in *Genocide Studies International.*

I write "obviously," but in fact it isn't always obvious. I'll finish this section by noting that I have read and reviewed many articles that inappropriately or unnecessarily included sections on study sites. Suffice it to say, a manuscript that examines the political economy of the Cambodian genocide probably does not need to include the longitude and latitude of Phnom Penh.

Coming to the End

In full disclosure, I find writing conclusions the least enjoyable part of the process. I'm not sure why; perhaps there's a psychological element of not wanting to finish. Then again, maybe I'm just anxious to finish and am already looking toward the next project. Regardless, of all the components of journal articles, I struggle most with conclusions.[75]

To begin (or, rather, to end): journal articles usually have either 'conclusions' or 'summaries.' This latter form may also appear as 'summary and findings.' Davidson-Arnott and Bauer finish their manuscript with "Conclusions"; Munro-Stasiuk and her co-authors end with "Discussion and Conclusions." The difference

---

Pre-Landsat 4 Satellite Imagery Using Vegetation Indices," *Remote Sensing* 11 (2019): 2397; and Alex R. Colucci, James A. Tyner, Mandy Munro-Stasiuk, Stian Rice, Sokvisal Kimsroy, Chhunly Chhay, and Corrine Coakley, "Critical Physical Geography and the Study of Genocide: Lessons from Cambodia," *Transactions of the Institute of British Geographers* 46, no. 3 (2021): 780–793.

75 For this reason, I find myself reading more and more the end-sections of manuscripts, in an effort to improve my writing techniques.

is subtle and, frankly, not always followed in practice. In general, 'summaries' should provide a *summation* of the article, that is, to restate however briefly the main thesis, the methods employed, and the principal findings or results. Conclusions do not (usually) summarize the entirety of the manuscript but instead provide a more expansive narrative of the article's contributions. Again, the difference is slight and, in practice, there is not really an essentialist forwarding of summaries or conclusions.

Regardless of section heading, there are key features that usually appear. The end-section should reaffirm the main thesis and the key takeaway. End-sections frequently make reference to broader contributions, or at least intimate how the study at hand can be extrapolated to consider other concepts, geographic locations, and so on. Often, writers will identify limitations to their work—although in many scientific formats, this discussion may warrant its own section. Writers may also identify key questions for future research.

Despite my aversion to writing conclusions, there is a key stylistic element that I really appreciate, namely when manuscripts effectively 'book-end' the introductory and concluding sections. This may seem obvious but, in my experience, many novice writers don't always follow through. Simply put, continuity is of the utmost importance when writing, but this holds especially for journal articles. In journal articles, there is a clear beginning and a clear ending. By way of contrast, encyclopedia entries are structured in a very different manner, similar to newspaper articles. In encyclopedia entries and news articles, for example, there is by convention a hierarchy in the presentation of material. The most important, declarative statement opens the piece, and subsequent material is presented in declining importance, relatively speaking. This doesn't mean *unimportant* but instead reflects that when journalists write newspaper stories, they don't readily know where the story will be cut off. Editors normally assign a certain amount of 'space' to the writer but, in the course of page mock-ups, stories may (and often are) cut earlier than planned.

Academic journal articles are structured differently. If a manuscript runs too long, the editor doesn't just lop off the concluding section. Instead, the author is asked to revise the manuscript, cutting words and sections throughout the whole of the manuscript. This follows because journal articles are structured as set-pieces with definitive beginnings and endings. Given this, it is preferable to make use of this 'bracketing' of the manuscript. On this point, if an author introduces a manuscript, say, by using a particularly distinctive metaphor, the concluding section should revisit this metaphor. I find that such an approach more effectively ties the manuscript together. And so, by way of illustration, I'll finish with an example of a bracketed introduction and conclusion from our collaborative work on Khmer Rouge irrigation schemes. The writing of this manuscript, to be clear, was a joint effort; the passages presented here, though, were crafted by Stian Rice.[76]

Our paper begins:

> A few kilometers downstream from Phnom Penh, on an artificial levee in the sedately meandering Bassak River, a team of pumping dredgers spit diesel soot and sputter red mud. Day and night, week after week, these rusting hulks toil, their labor unbroken until monsoonal floods force a temporary retreat to higher ground. Later, as the waters recede, the heavy equipment returns to the levee to begin again, collecting the precious material carried and deposited here from other parts of the Mekong Basin. Months from now, the sand and gravel these dredgers pull from the Bassak will follow the meandering channels of global capital to Singapore, where they will anchor new waves of accumulation.
>
> This movement of earth is not the first time Cambodia's riparian and lacustrine sediments have been enlisted into the service of capital. The country's central plain is one of the largest and flattest basins in Southeast Asia: a 25,000 sq km expanse surrounding Cambodia's great lake, the Tonle Sap. ... In the 1970s, the systematic movement of Cambodian earth played a key role in the economic plans of the Communist Party of Kampuchea (CPK), otherwise known as the Khmer Rouge. These plans precipitated one of the bloodiest genocides of the 20th century. In fact, just a few kilometers to the north-west of today's Bassak River dredgers is Choeung Ek, the site of mass graves from the Khmer Rouge period and now a genocide memorial and tourist attraction. ...
>
> If Cambodian earth could speak, these sediments would convey a long history of being broken up, lifted, moved, compacted, inundated, and baked under the sun. For centuries, Cambodian earth—built by water—has been enlisted to hold water back, direct its flow, and absorb its excess. These soils have sustained delicate rice roots and with these plants ensured the survival of millions. This ground was both the discursive and material foundation for the CPK's infamous irrigation plans. And today, Cambodian earth still holds in its embrace the remains of millions, many of whom were tamped down into the same earthworks they helped to build. To understand this genocide, then, is to understand the CPK's plans for rice, their plans to irrigate that rice, and their efforts to violently inscribe that irrigation into the surface of the earth.[77]

Our conclusion returns both to the imagery and the tone established from the beginning. Notably, given the structure of the manuscript, key takeaways are presented in the penultimate section. This allows us the opportunity to more effectively—aesthetically—bracket the paper. We finish:

> Forty-five years after the CPK broke ground on Trapeang Thma Dam, Cambodia continues to transform its sand, gravel, and laterite soils in the interest of capital accumulation. Foreign investment has jump-started construction of large-scale irrigation schemes, many a revival of CPK-era plans. Intensive rice production for export is still a centerpiece of the Cambodian

---

76 Colucci et al., "Critical Physical Geography"
77 Colucci et al., "Critical Physical Geography," 781–782.

economy, and with it, industrial technologies that automate the work of plough and ox, and reconstitute the soil through chemical inputs. Hydroelectric projects along the Mekong threaten to inundate upstream valleys and deprive downstream agriculture of fertile sediments. And today, some 20 million cubic meters of sand and gravel leave Cambodia each year; not suspended in the flow of the Mekong and Bassak Rivers but suspended in barges bound for Singapore. This grand diaspora of dirt is but the latest geophysical transformation to shape the country's watery landscape, its economy, and its people. It remains to be seen what legacies of precarity and violence this transformation will leave. In this paper, we applied the lens of critical physical geography to explore the role of geophysical and hydrosocial transformation in the production of genocide. From this perspective, we find not only that Cambodian earth can speak, it also has a great deal to say. The epistemological approach inherent to critical physical geography affords us a means of listening.[78]

The Heart of the Matter

Thus far, we've considered the structure of journal articles and have provided some insight into the writing of key sections, notably introductions, methods sections, study sites, and conclusions. Still unresolved is the weighty matter of substance. Stated differently, we've established the skeleton of journal articles. What's left is to add the flesh and muscle of the manuscript.[79] We need, at this point, to consider three overarching elements that give body to the manuscript: *substance, balance*, and *rigor.*

Let's start with the substance of an article. Here, I use substance in reference to the overall composition of the manuscript. We've already considered some aspects of the substantive make-up of a journal article. Do we or do we not include sections on methodologies, for example? Substance though entails *something more.* Perhaps an analogy will help. We can think of various 'pieces' of research, for example previous literature, concepts, data collection procedures, analytic techniques, and so on, as ingredients. However, as anyone who has cooked will immediately understand, cooking is more than the compilation of ingredients. When we bake cookies, we often mix the dry ingredients (e.g., flour, salt, baking powder) separately from the 'wet' ingredients (e.g., eggs, butter, sugar, vanilla). Moreover, how these ingredients are blended together matters greatly. Depending on the way we combine the ingredients, our cookies might be too flat or too thick, too brittle or too chewy. The same holds with writing a journal article. It isn't just a matter of including or not including any par-

---

78 Colucci et al., "Critical Physical Geography," 790–791.
79 I am indebted to Don Mitchell who, having reviewed an earlier draft of this chapter, pushed me to develop this section in greater detail.

ticular section. The end-product, similar to a cookie, is more than a list of ingredients. Ultimately, at this point we're asking how the overall manuscript holds together. This relates to a second element, balance.

We've seen that there is no standard template, or outline, for journal articles. Manuscripts might follow a more conventional structure: introduction, literature review, methods, study site, results and discussion, and conclusion. Conversely, manuscripts might be more free formed, perhaps with two or three 'conceptual' sections. Regardless, once you have a sense of your intended structure you can start to plan approximate word counts for each section.[80] If you follow a traditional format, the following word counts can serve as a useful rule of thumb:

1. Introduction (500–750 words)[81]
2. Literature review (750–1,000 words)
3. Methods (750 words)
4. Findings (1,500 words)
5. Discussion (1,500 words)
6. Conclusion (500–750 words)

In this example, the final word count will run between 5,000 and 7,000 words—a good target range for most journals. Remember that some journals, such as the *Annals of the American Association of Geographers*, permit longer manuscripts, in excess of 10,000 words, while other journals, for example *The Professional Geographer* and *Area*, limit manuscripts to about 5,000 words. In addition, in many journals, bibliographies/references and figures, maps, and tables are often included in the final word count.

Although these word counts should serve as a guide only, there's some additional points worth considering. As I explained earlier, I think journal articles are more effective when the introduction and conclusion work as book-ends. This holds, therefore, in the size of these two sections. Manuscripts can appear unbalanced, for example, if your introduction rambles on for 1,500 words but ends abruptly, after, say, 500 words. Likewise, something seems off (usually) if your literature review (if included) runs on for 2,000 words but your empirical findings come in at a paltry 750 words.

Lastly, we need to consider the rigor of the manuscript. How do we marshal the 'evidence' in our manuscript, whether this is understood as 'empirical find-

---

80 A useful discussion is provided in Linda Finlay, "How to Write a Journal Article: Top Tips for the Novice Writer," *European Journal for Qualitative Research in Psychotherapy* 10 (2020): 28–40; at 34.

81 It isn't always easy for novice writers to think in terms of words. Roughly speaking, one 8 ½ × 11 inch page of double-spaced text written in 12-point font translates into 300 words.

ings' or 'theory'? To a large degree, this is the most challenging element of a journal article (or, frankly, any piece of academic writing), and yet is the most difficult to convey in any concise manner. Indeed, it is frustrating that many otherwise helpful publications on writing journal articles skip entirely this element. In his chapter on writing and publishing in human geography, for example, Larry Bourne includes valuable lessons on "Starting Out", "Selecting Journals", "Reading Your Audience", "Rejection", and "Collaborative Research".[82] In the all-important section on 'crafting the article,' however, there's not much meat. In part, as Bourne correctly identifies, "the task of designing an article for publication is essentially an individualized exercise."[83] However, beyond a recantation of fairly conventional design elements, one is hard-pressed to really address how the overall argument or thesis is made. It may seem obvious, Kate Turabian writes, that "you must back up a claim with reasons and evidence, but it's easy to confuse those two words because we often use these as if they mean the same thing."[84] In other words, journal articles *in general* require both reason and evidence in support of the manuscript's objective. This holds even for review essays and progress reports, for example those published in *Geography Compass* or *Progress in Human Geography.*

Let me hasten to add, however, that not all manuscripts require 'empirical' evidence, if by that you think of numbers and statistics. Both qualitative and quantitative methods have their own criteria on which to evaluate the 'robustness' of research. So-called quantitative research, for example, is often judged on such considerations as validity, replicability, and reliability. Qualitative research is also evaluated for its rigor, but on different terms. Indeed, as Matt Bradshaw and Elaine Stratford underscore, rigor is vitally important in qualitative research and this should be apparent in the representation of one's research.[85] To that end, several strategies exist to demonstrate rigor in qualitative research: verbatim respondent quotations, details of interview practices, discussions of the procedures for analysis, immersion/lengthy field work, revisits to respondents, verification by respondents, appeals to interpretative communi-

---

82 Bourne, "On Writing and Publishing."
83 Bourne, "On Writing and Publishing," 102.
84 Kate L. Turabian, *A Manual for Writers of Research Papers, Theses, and Dissertations*, 7th edition (Chicago: University of Chicago Press, 2003), 52.
85 Matt Bradshaw and Elaine Stratford, "Qualitative Research Design and Rigour," in *Qualitative Research Methods in Human Geography*, edited by Iain Hay (Oxford: Oxford University Press, 2005), 67–76.

ties, and the provision of a rationale for verification of the findings.[86] Building on these suggestions, Cathy Bailey, Catherine White, and Rachel Pain suggest nine principles for the evaluation of qualitative research in human geography: the need for theoretical sensitivity; reflexive management that strengthens qualitative validity; constant comparison by continued questioning; thorough documentation of procedures to leave a 'paper trail' audit that strengthens qualitative validity; clear and open reporting of procedures; clear discussion of how theory 'fits' the reality of the respondents' lives—with rationale offered for 'negative' cases; generation of criteria for evaluation of particular research; recognition of the researcher(s) influence on the research findings; and use of archives for data and documentation relating to research procedures.[87]

More recently, several geographers, such as Sarah De Leeuw, Dydia DeLyser, Harriet Hawkins, and Clare Madge, among others, have advocated for more creative approaches to the doing and writing of research.[88] Notable is the premise that the conduct of research is inseparable from the writing process; in other words, there is no 'gap' or 'separation' between the planning and execution of writing and the planning, design, writing, and revising of manuscripts. This holds tremendous importance, therefore, in how the 'final' manuscript appears. Indeed, one might choose not to write a journal article in any conventional sense.

Some Final Thoughts on Writing Journal Articles

In the introductory chapter I provided a brief discussion of my writing practices. Here, I want to conclude with some thoughts related specifically to writing journal articles for an academic audience. There is, of course, no one-size-fits-all model for journal articles; there is no universal template to structure the manuscript nor any fail-safe criteria or principles which will guarantee substance and rigor to our articles. Yes, conventions exist, both within geography's sub-fields and for specific

---

86 Jamie Baxter and John Eyles, "Evaluating Qualitative Research in Social Geography: Establishing 'Rigour' in Interview Analysis," *Transactions of the Institute of British Geographers* 22, no. 4 (1997): 505–525; at 508.

87 Cathy Bailey, Catherine White, and Rachel Pain, "Evaluating Qualitative Research: Dealing with the Tension between 'Science' and 'Creativity,'" *Area* 31, no. 2 (1999): 169–183; at 175.

88 Dydia DeLyser and Harriet Hawkins, "Introduction: Writing Creatively—Process, Practice, and Product," *Cultural Geographies* 21, no. 1 (2014): 131–134; Clare Madge, "On the Creative (Re)turn to Geography: Poetry, Politics and Passion," *Area* 46, no. 2 (2014): 178–185; and Sarah De Leeuw and Harriet Hawkins, "Critical Geographies and Geography's Creative Re/turn: Poetics and Practices for New Disciplinary Spaces," *Gender, Place & Culture* 24, no. 3 (2017): 303–324.

journals. Different writers, likewise, often exhibit a preference toward one writing style as opposed to another. And still other geographers challenge the entire premise of writing refereed journal articles.

So where does this leave us? I conclude with some basic suggestions, keeping in mind the different normative traditions that exist in geography. To be effective (and to not be rejected!) the form and format, tone and style must align with the guidelines specified by any given journal. What works in *Gender Place and Culture* may not work in *Quaternary Science Reviews*, and vice versa. However, certain elements remain constant. Writing should effectively communicate the subject-matter, whether it is predominantly empirical or theoretical. There should be continuity and cohesion within the manuscript; that is, the separate sub-sections are not disparate but comprise a whole. We're making cookies, remember, not just throwing together a bunch of ingredients.

Chapter 4 Book Chapters

Let's start with the obvious. You write book chapters for edited books. And, on the face of it, book chapters resemble journal articles. Book chapters and journal articles are often of similar length, typically between 5,000 and 8,000 words. The structure is often very similar, with stand-alone sections separated by sub-headings. So, isn't writing a book chapter comparable to writing a journal article? Can't the same general principles of writing articles hold for writing chapters? Actually, no. But this is to get ahead of the story. Before we discuss *how* to write a book chapter—and how this differs from writing a journal article—we need to ask ourselves; why do edited books even exist?

According to Ron Martin, "of all the different forms of research and writing, producing an edited book is often regarded as the least significant type of academic activity, and certainly as having the lowest priority in building an academic career."[1] This perception, whether warranted or not, largely holds for geography. The reasons are many: edited books (and the chapters they contain) are not peer reviewed—or at least, not to the same standards as refereed journal articles; editor(s) receive the credit for a publication based on the efforts of others (the contributors); edited books (and the chapters that have been written for them) lack the potential impact and visibility of authored books and journal articles; time spent on putting edited books together could be better directed toward more 'valuable' projects, such as writing refereed journal articles.[2] Certainly, we can quibble with all of these reasons. Many edited books do undergo extensive peer review, comparable in standard to reputable journals. Also, many chapters in edited books, if not the book itself, can have a sizable impact on the field. Indeed, it is not unheard of for some chapters to garner considerably more citations than refereed articles.[3] Indeed, as Martin defends, "just as there are indifferent and low-impact edited books, so too there are many indifferent and low-impact authored books and innumerable instantly forgettable journal articles."[4]

Arguments to the contrary, the fact remains that "particularly within the circles of academic promotion boards, staff appointment panels and research assess-

---

1 Ron Martin, "Why Edit Books? In Defense of an Oft-Disparaged Academic Activity," *Regional Studies* 47, no. 9 (2013): 1611–1613; at 1611.

2 Martin, "Why Edit Books?" 1611.

3 I'm not necessarily advocating that citation counts are the determining factor of a manuscript's 'worthiness'. For now, however, the reality is: citations matter in the neoliberal university. We'd be foolish to think otherwise, and so we need to acknowledge this factor.

4 Martin, "Why Edit Books?" 1611.

https://doi.org/10.1515/9783111189727-006

ment bodies ... edited books do not count for much."[5] By corollary, unfortunately, book chapters that appear in edited books also do not count for much. In my experience, having served on a number of internal and external hiring committees and tenure/promotion committees, edited books and book chapters are routinely seen as being of lesser importance. And on those few occasions I edited books, it was a perennial challenge to solicit scholars to contribute a chapter. Often, prospective contributors would indicate their interest and support of the project, but decline because their time was best served writing journal articles.

Edited books—and chapters appearing in edited books—do have much to contribute. To appreciate this assertion, though, requires a closer look at the functions provided by edited books. In so doing, we can better understand how to (and why) write book chapters in edited books.

Why Edit Books?

The case for, and value of, edited books, Martin argues, is two-fold. On the one hand, an edited collection of original, cutting-edge essays can have a significant impact on the intellectual development of a discipline. By bringing together the leading scholars in a field around a common objective—by venturing new ideas and setting out a possible future research agenda—edited books can actually help to establish new sub-fields or specialisms.[6] For example, *Places Through the Body*, edited by Heidi Nast and Steve Pile, and *Thresholds in Feminist Geography*, edited by John Paul Jones, Heidi Nast, and Susan Roberts, are foundational texts in the development of feminist geography; *Black Geographies and the Politics of Place*, edited by Katherine McKittrick and Clyde Woods, is central to the establishment of Black geographies; and *Place, Power, Situation, and Spectacle*, edited by Stuart Aitken and Leo Zonn, is a keystone of film geographies.[7]

On the other hand, edited books provide an important community function within the discipline. Many graduate students and early-career faculty express

---

5 Martin, "Why Edit Books?" 1611.
6 Martin, "Why Edit Books?" 1612.
7 Stuart C. Aitken and Leo E. Zonn (eds), *Place, Power, Situation, and Spectacle: A Geography of Film* (Lanham, MD: Rowman & Littlefield, 1994); John Paul Jones, III, Heidi J. Nast, and Susan M. Roberts (eds), *Thresholds in Feminist Geography: Difference, Methodology, Representation* (Lanham, MD: Rowman & Littlefield, 1997); Heidi J. Nast and Steve Pile (eds), *Places Through the Body* (New York: Routledge, 1998); and Katherine McKittrick and Clyde Woods (eds), *Black Geographies and the Politics of Place* (Cambridge, MA: South End Press, 2007).

feelings of isolation.[8] In a recent study, for example, Roberta Hawkins, Maya Manzi, and Diana Ojeda find that many graduate students "spoke of wanting to engage in more collaborative work, especially across perceived divisions in political, theoretical and methodological approaches, but concluded that the pressures in their program or the entrenchments of fields of geography discouraged this."[9] Participation in collaborative projects, including edited books, *may* provide some respite.[10] Martin suggests that editing books, and contributing to edited books, contributes to academic community building: collaboration and cooperation in these projects help establish networks for the development and diffusion of ideas, but also help promote a sense of togetherness. If done well, Martin concludes, the project "can foster interchange between all those involved."[11]

Edited books do have a place in geography. Let's delve a bit more deeply into the 'right' and 'wrong' reasons to edit a book. Mark Davis and Bernd Blossey identify several reasons why *not* to produce an edited book, including:

1. It is easier than writing a book yourself and you still get to list it as a book on your CV.
2. It is safer than writing one. While edited books seldom get rave reviews, they are also seldom publicly renounced.

8 Michael N. Solem and Kenneth E. Foote, "Concerns, Attitudes, and Abilities of Early-Career Geography Faculty," *Annals of the Association of American Geographers* 94, no. 4 (2004): 889–912; Kenneth E. Foote, "Creating a Community of Support for Graduate Students and Early Career Academics," *Journal of Geography in Higher Education* 34, no. 1 (2010): 7–19; Roberta Hawkins, Maya Manzi, and Diana Ojeda, "Lives in the Making: Power, Academia and the Everyday," *ACME: An International Journal for Critical Geographies* 13, no. 2 (2014): 328–351; Alison Mountz, Anne Bonds, Becky Mansfield, Jenna Loyd, Jennifer Hyndman, and Margaret Walton-Roberts, "For Slow Scholarship: A Feminist Politics of Resistance through Collective Action in the Neoliberal University," *ACME: An International Journal for Critical Geographies* 14, no. 4 (2015): 1235–1259; Alison Mountz, "Women on the Edge: Workplace Stress at Universities in North America," *The Canadian Geographer/Le Géographe canadien* 60, no. 2 (2016): 205–218; Maya Manzi, Diana Ojeda, and Roberta Hawkins, "'Enough Wandering Around!': Life Trajectories, Mobility, and Place Making in Neoliberal Academia," *The Professional Geographer* 71, no. 2 (2019): 355–363; and Lydia Wood, L. Kate Swanson, and Donald E. Colley III, "Tenets for a Radical Care Ethics in Geography," *ACME: An International Journal for Critical Geographies* 19, no. 2 (2020): 424–447.
9 Hawkins et al., "Lives in the Making," 337.
10 Admittedly, for many precarious faculty, writing manuscripts—including book chapters for edited books—is part of the problem. Overburdened and contingently employed, these writers struggle to find the time, resources, and support. Unfortunately, owing to the negative perception of publishing in edited books, these scholars may be compelled to *not* participate, devoting instead their efforts toward the more praiseworthy refereed article. It can become a vicious circle.
11 Martin, "Why Edit Books?" 1612–1613.

3. Writing a chapter is an easy way for you and your friends to get an additional publication without having to go through the rigors of real peer review.
4. You want to provide the field with a collection of related writings on a topic that is already saturated in the primary literature.
5. A publisher approaches you to edit a book and you agree without first determining that there is a need for it.[12]

Clearly, these 'wrong' reasons help fuel the bad reputation held by many scholars throughout academia. So, what are the right reasons to edit a book, apart from the epistemic and social reasons offered by Martin? Davis and Blossey provide three:

1. You want to provide the field with a collection of related writings on a new area of research or inquiry.
2. You want to form a new synthesis, for example by bringing together diverging fields that live separate disciplinary lives in specialty journals.
3. You want to provide a complete overview of a topic and invite leading specialists to contribute.[13]

With an understanding of why edited books come into existence, we can now say something about the process of writing chapters for edited books. As we'll see, many of the suggestions offered in my chapter on journal articles hold true. That said, there are some differences worth exploring.

Participating in an Edited Book Project

Edited books are at their best, Davis and Blossey write, "when they provide a well-integrated collection of chapters around a particular theme that represents a new area of research or inquiry or a novel synthesis (that may challenge existing common ground or dogma)."[14] Consequently, editors often establish one or two specific objectives for the book and work to ensure that all authors write their chapters to align with the overarching vision of the project.[15] This does not imply that authors must slavishly 'buy into' a dominant interpretation or approach; indeed, contributors might provide much-needed counter-arguments. In the edited book *Spatiality,*

---

12 Mark A. Davis and Bernd Blossey, "Edited Books: The Good, the Bad, and the Ugly," *Bulletin of the Ecological Society of America* 92, no. 3 (2011): 247–250; at 249.

13 Davis and Blossey, "Edited Books," 249.

14 Davis and Blossey, "Edited Books," 247.

15 Davis and Blossey, "Edited Books," 250.

Sovereignty and Carl Schmitt, for example, Stephen Legg assembled a group of scholars to consider the legacy and significance of the controversial German jurist Carl Schmitt.[16] Some contributors wrote positively of Schmitt's importance, others were considerably less generous.

When editors seek to consolidate a recently developed sub-field or to bring together scholars working on a newly emerging area of research, they will ask contributors to position their chapters accordingly.[17] In other words, the proposed volume is more than a collection of chapters but instead forms a coherent body of scholarship. And while an edited volume can seldom achieve the same level of analysis, synthesis, and coherence as a monograph, the end-product can be no less pathbreaking.[18] This point is extremely important when drafting book chapters. Whereas journals provide 'guidelines' on style and substance, editors (frequently) impose an overarching conceptual framework for the book prior to inviting colleagues to contribute.[19] In *Economics of Death: Economic Logics of Killable Life and Grievable Death*, editors Patricia Lopez and Kathryn Gillespie ask contributing authors to curate "a dialogue across disciplines in order to reach beyond geography and engage a more holistic vision of how economies of death operate in the world."[20] In other words, Lopez and Gillespie encouraged and required writers to consider how the completed edited book would be more than the sum of its parts. Earl Harper and Doug Specht adopted a similar approach in their edited book, *Imagining Apocalyptic Politics in the Anthropocene*.[21] For this project, Harper and Specht called on contributors to respond to several questions centered on contemporary apocalyptic events such as climate change and micro-plastic pollution.

Not all edited books are efforts to establish or respond to newly emerging areas of research. Some edited books attempt to provide clarity to existing fields, for example by providing overviews of trends. Frequently, chapters in these edited books are composed almost entirely of secondary sources and hold much in common with review essays found in journals. In *Geography in America at the Dawn of the 21st Century*, edited by Gary Gaile and Cort Willmott, teams of scholars repre-

---

16 Stephen Legg (ed.), *Spatiality, Sovereignty and Carl Schmitt: Geographies of the Nomos* (New York: Routledge, 2011).

17 Martin, "Why Edit Books?" 1612.

18 Davis and Blossey, "Edited Books," 250; Martin, "Why Edit Books?" 1612.

19 Davis and Blossey, "Edited Books," 250.

20 Kathryn A. Gillespie and Patricia J. Lopez, "Introducing Economies of Death," in *Economies of Death: Economic Logics of Killable Life and Grievable Death*, edited by Patricia J. Lopez and Kathryn A. Gillespie (New York: Routledge, 2015), 1–13; at 4.

21 Earl T. Harper and Doug Specht (eds), *Imagining Apocalyptic Politics in the Anthropocene* (New York: Routledge, 2022).

senting the many specialty groups of the American Association of Geographers were assembled to provide surveys of their respective sub-fields.[22] Other edited books are designed as text-books, for use in specialized courses, for example *Cities of the World*, edited by Stanley Brunn, Maureen Hays-Mitchell, Donald Zeigler, and Jessica Graybill.[23]

Writing for an Edited Book

When writing journal articles, writers often begin the process by thinking through the constituent elements required, for example the theoretical or conceptual framing, the broader literature, and so on. When writing book chapters, much of this planning has already been provided by the editors. This does not mean that authors have no 'free will' in their writing. Instead, authors are able to exclude some material that otherwise would be included in a journal article. Some examples should clarify this point.

In *After Heritage: Critical Perspectives on Heritage from Below*, Hamzah Muzaini and Claudio Minca ask contributors to address the concept of 'heritage from below.'[24] However, in their introduction Muzaini and Minca provide an in-depth discussion of relevant terms and concepts; they also provide an historical overview of heritage studies. This provides theoretical and conceptual coherence to the edited book and also grounds the contributions within a common scholarly lineage. In turn, contributors are able to concentrate more fully on their case studies and how these address the project as a whole. On the other hand, in *Black Geographies and the Politics of Place*, editors Katherine McKittrick and Clyde Woods want to provide a forum to "initiate a discussion of how we might begin to work through the dilemmas that continually come forth when race and space converge with one another and relegate black geographies to bodily, economic/historical materialist, or metaphoric categories of analysis."[25] They explain, "The articles collected here are interdisciplinary discussions

---

22 Gary L. Gaile and Cort J. Willmott (eds), *Geography in America at the Dawn of the 21st Century* (Oxford: Oxford University Press, 2003).

23 Stanley D. Brunn, Maureen Hays-Mitchell, Donald J. Zeigler, and Jessica K. Graybill, *Cities of the World: Regional Patterns and Urban Environments*, 6th edition (Lanham, MD: Rowman & Littlefield, 2016).

24 Hamzah Muzaini and Claudio Minca (eds), *After Heritage: Critical Perspectives on Heritage from Below* (Northampton, MA: Edward Elgar, 2018).

25 Katherine McKittrick and Clyde Woods, "'No One Knows the Mysteries at the Bottom of the Ocean,'" in *Black Geographies and the Politics of Place*, edited by Katherine McKittrick and Clyde Woods (Cambridge, MA: South End Press, 2007), 1–13; at 6.

of race and space, with specific reference to black geographies."[26] Meaningfully, McKittrick and Woods clarify, "Each of the chapters here can be read separately, or together with another chapter, or in any order. While the central focus of each discussion is geography—as it is articulated through the physical, imaginary, and political concerns of black diasporic subjects—the writing is not meant to replace or identify the limits of existing debates in human geography."[27] Compared with the parameters established in Muzaini and Minca's edited volume, the format for McKittrick and Woods' volume is broader in scope, meant to stimulate debate precisely through the multiple readings of the subject.

As a final example, Scott Kirsch and Colin Flint's edited volume *Reconstructing Conflict: Integrating War and Post-War Geographies* occupies a middle position. From the outset, they establish a set of guiding concepts and debates but call upon chapter authors to challenge and expand upon these.[28] In their introduction, Kirsch and Flint write, "The continuity of processes across what are commonly termed conflict/post-conflict situations is a matter of the on-going power relations that contest the way spaces and place are made, maintained, and altered."[29] This establishes the immediate purpose of the edited collection. They explain:

> To delve further into these worlds, and to expand our effort to conceptually integrate war and post-war geographies, there is a need to think across diverse contexts. The chapters collected in this volume provide such geographical breadth, including studies encompassing twentieth and early twenty-first century landscapes of war and reconstruction extending from East and Southeast Asia to the Middle East to Europe and North America. The diverse regional expertise of the contributors, and the varied theoretical perspectives and methodological approaches embodied in their work, offer a wide scope onto processes of conflict and post-conflict reconstruction. While each chapter in some ways stands alone, each has been commissioned and is intended to contribute to the broader goals sketched here, and to encourage readers, whether researchers, students, practitioners, or general readers, to learn and think critically about the intersections of war, peace, and reconstruction processes. Our aim in putting the book together has thus been, collectively, to develop new geographical approaches for understanding how spaces of war have been reconstructed as ostensible zones of peace, but with their own dynamics of conflict.[30]

---

26 McKittrick and Woods, "'No One Knows,'" 7.
27 McKittrick and Woods, "'No One Knows,'" 8.
28 Scott Kirsch and Colin Flint (eds), *Reconstructing Conflict: Integrating War and Post-War Geographies* (Burlington, VT: Ashgate, 2011).
29 Scott Kirsch and Colin Flint, "Introduction: Reconstruction and the Worlds that War Makes," in *Reconstructing Conflict: Integrating War and Post-War Geographies*, edited by Scott Kirsch and Colin Flint (Burlington, VT: Ashgate, 2011), 3–28; at 19–20.
30 Kirsch and Flint, "Introduction," 20.

My takeaway is that when writing for an edited book, authors enact a different pre-writing process, one that is (often) highly circumscribed by the editor(s). Likewise, both the drafting and editing of the manuscript are dissimilar from those of when writing journal articles. In edited books, editors will typically assume a more hands on approach to ensure coherence and continuity within and between individual chapters. When writing journal articles, you have considerable leeway to determine the style and substance of your manuscript; you make the decision—at least, prior to the review process—to include or exclude sections. When writing book chapters, many of these decisions are out of your hands. Certainly, you can (and should) work closely with the editor(s) in planning and drafting your chapter; however, in the end, your chapter is more a collaboration than an individual contribution.

The Structure of Book Chapters in Edited Volumes

Many journal articles adopt a fairly traditional format: Introduction, Literature Review, Methods, Findings, and Conclusions. Book chapters may, and often do, depart remarkably from this format. By way of illustration, let's look at the outlines of two chapters from McKittrick and Woods' *Black Geographies*. In doing so, recall McKittrick and Woods' motivation in compiling this edited collection: to provide "interdisciplinary discussions of race and space, with specific references to black geographies" in order to "better understand the racialization that has long formed the underpinnings for the production of space."[31] So framed, let's see how Peter Hudson structured his contribution. His chapter, "'The Lost Tribe of a Lost Tribe': Black British Columbia and the Poetics of Space," consists of five sections:

1. [Untitled introduction]
2. Ethel Wilson and the Conditions of Blackness
3. The Imagining of Joe Fortes
4. The 'Different Worlds' of the 'Innocent Traveller'
5. Negotiating Spaces and Places[32]

Clearly, Hudson does not follow a conventional, overly rigid format as is typical in many journal articles. There is no stand-alone section on theory or method, for example; nor is there an obvious 'literature review' section. This does not imply, though, that Hudson's contribution is any less scholarly. In his chapter, Hudson

---

31 McKittrick and Woods, "'No One Knows,'" 7–8.
32 Peter James Hudson, "'The Lost Tribe of a Lost Tribe': Black British Columbia and the Poetics of Space," in *Black Geographies and the Politics of Place*, edited by Katherine McKittrick and Clyde Woods (Cambridge, MA: South End Press, 2007), 154–176.

provides a narrative and this resonates with the overall tenor of the edited volume. He sets the scene, so to speak, by way of introducing the notion of Black geographies in British Columbia. Hudson writes, "Black people live there, so I am told, and the proliferation (if not deluge) of black writing from the province over the past decade seems to both prove this basic residential contention while simultaneously demonstrating the insisting value of folk knowledge."[33] That said, Hudson's chapter aligns with McKittrick and Woods' desire to understand geographies "not as distinct areas or nations, but as overlapping diaspora spaces."[34]

A second example is Clyde Woods' "'Sittin' on Top of the World': The Challenges of Blues and Hip Hop Geography." It also is composed of five sections and is seemingly lacking in structure:

1. [Untitled introduction]
2. "Standing Here Lookin' One Thousand Miles Away": Sites for the Production of Geographic Knowledge
3. "Yuh Can Read My Letters but Yuh Sho Cain't Read My Mind": The Origins of Blues Geographic Knowledge
4. Hip Hop as a Blues Movement
5. House of the Blues: The Structures of Geographic Knowledge[35]

In his contribution, Woods argues that "the blues [in addition to being a musical tradition] is [also] a knowledge system indigenous to the United States that is expressed through an ever-expanding variety of cultural, economic, political, and social traditions."[36] In building his argument, Woods juxtaposes the music of the blues with human geography and demonstrates that "Embedded within the blues tradition are highly developed and institutionalized forms of philosophy, political economy, social theory and practice, and geographic knowledge that are dedicated to the realization of global social justice."[37] The organization of Woods' chapter itself is lyrical, and the structure of the chapter appears not as a strict framework that circumscribes the narrative but instead reads as a graceful composition.

Both Hudson and Woods' chapters are powerful as stand-alone pieces. However, the importance of their narratives is augmented when read together. And this is

---

33 Hudson, "'The Lost Tribe,'" 154.
34 McKittrick and Woods, "'No One Knows,'" 8.
35 Clyde Woods, "'Sittin' on Top of the World': The Challenges of Blues and Hip Hop Geography," in *Black Geographies and the Politics of Place*, edited by Katherine McKittrick and Clyde Woods (Cambridge, MA: South End Press, 2007), 46–81.
36 Woods, "'Sittin' on Top of the World,'" 49.
37 Woods, "'Sittin' on Top of the World,'" 49.

the potential strength of successful edited books. Journal articles, on their own, can and should build on existing scholarship; this is of course the point of 'literature reviews'. However, journal articles are (typically) planned and published as autonomous pieces of scholarship. Book chapters, with few exceptions, are frequently situated within a particular literature, chiefly the companion chapters that comprise the edited book.

What about writing for edited books geared toward pedagogical purposes? As a type of text-book, edited books often require an even greater level of structure than that found in more research-oriented volumes. *Contemporary Ethic Geographies in America*, co-edited by Ines Miyares and Christopher Airriess, provides a useful example.[38] Unlike *Black Geographies*, Miyares and Airriess' edited book is designed and compiled specifically for use in the classroom. This often requires a greater standardization of chapters. Here's the outline for Marie Price's chapter "Andean South Americans and Cultural Networks":

1. [Untitled introduction]
2. Andean Migration to the United States
3. Settlement and Distribution Patterns
4. Social and Economic Attainment
5. Community Case Study: Bolivians in Washington
6. Life in Metropolitan America
7. Transnational Networks
8. Conclusion[39]

And here's Wei Li's outline for her contribution on Chinese Americans:
1. [Untitled introduction]
2. Immigration History
3. Settlement Patterns
4. Globalization and Contemporary Chinese American Communities
5. Socioeconomic Characteristics
6. A Thriving Ethnic Economy
7. Transnational Connections
8. Ethnic Identity and Racialized Challenges

---

38 Ines M. Miyares and Christopher A. Airriess (eds), *Contemporary Ethnic Geographies in America* (Lanham, MD: Rowman & Littlefield, 2007).
39 Marie Price, "Andean South Americans and Cultural Networks," in *Contemporary Ethnic Geographies in America*, edited by Ines M. Miyares and Christopher A. Airriess (Lanham, MD: Rowman & Littlefield, 2007), 191–211.

9. Conclusion[40]

Lastly, here's the outline for my contribution on Filipinos in the United States:
1. [Untitled introduction]
2. Context of Entry
3. Settlement Patterns
4. Immigrant Experiences
5. The Forgotten Immigrants
6. The Pan-Nationalism of the Philippine Diaspora
7. Conclusion[41]

During the planning stage for this project, Miyares and Airriess asked contributors to follow a pre-set format. Writers were required to address specific topics, including, for example, the context of entry (immigration history), settlement patterns, and transnational connections. In addition, we were expected to provide a case study, to capture something of the lived experience of immigrants in the United States. As such, most authors adopted a fairly similar structure for their chapters. We were given a modicum of leeway in organization, and for any given chapter exceptions were made based on the particularities of the case study.

Some Final Thoughts on Writing Book Chapters

There is a high degree of affinity between journal articles and book chapters. It is a mistake, however, to presume that writing a book chapter is no different from writing an article. The editorial oversight is often more intense and takes a more substantive form. Book editors frequently demand a degree of cohesion and continuity both within and between chapters; as such, individual writers are somewhat circumscribed in the tone, style, and structure of their writing. This is not a universal rule, of course. Some editors extend significant flexibility to their contributors. That said, book chapters are ultimately *collaborative projects* that are co-produced by the editor and the writer. This alone sets book chapters apart from journal articles in ways unexpected to many novice writers.

---

40 Wei Li, "Chinese Americans: Community Formation in Time and Space," in *Contemporary Ethnic Geographies in America*, edited by Ines M. Miyares and Christopher A. Airriess (Lanham, MD: Rowman & Littlefield, 2007), 213–232.
41 James A. Tyner, "Filipinos: The Invisible Ethnic Community," in *Contemporary Ethnic Geographies in America*, edited by Ines M. Miyares and Christopher A. Airriess (Lanham, MD: Rowman & Littlefield, 2007), 251–270.

Chapter 5 Scholarly Monographs

Trevor Barnes, in a recent essay appearing in *Progress in Human Geography*, writes about his experiences serving on the Faculty of Arts Promotion and Tenure Committee at the University of British Columbia.[1] From his experiences, Barnes relates, he gained valued insight into the different standards for tenure and promotion across the various disciplines. In psychology, for example, the ticket to tenure and promotion is the multi-authored scientific paper with trip-figure citations; in history and English, it is the single-authored book published with an appropriate (i.e., reputable) scholarly press.

At Kent State University, I have served on a comparable college-level committee for many years. And my experience has pretty much echoed that of Barnes. Faculty employed in the so-called natural or hard sciences place heavy weight on refereed journal articles in top-tiered journals, complemented with funding from external agencies, for example the National Science Foundation and the National Institute of Health. In history, faculty are expected to have published a solo-authored research monograph for tenure and promotion to associate professor; a second book is required for promotion to professor. In a way, tenure and promotion decisions for *some* faculty in the humanities are simple: either you've published a research monograph, or you haven't. The problem that arises most for these cases is the definition of a research monograph. Strange as it may seem to some, not all books count.

This chapter isn't the place to define precisely and conclusively what is or is not an acceptable research monograph. Experience has taught me that this is often a political matter, with faculty passionately debating their position against that of their colleagues.[2] Admittedly, there is a degree of snobbery involved: some presses are reputed to be more prestigious. There is the matter, also, of length and contribution. In modern and classical languages, for example, some faculty labor for years translating a novel or some other piece of writing; they provide an in-depth, historically grounded context of the manuscript, coupled with detailed description of 'translator notes.' In its totality, does this constitute a 'scholarly monograph'? Again, it isn't my purpose to adjudicate what is or is not an academic

---

1 See Trevor Barnes' essay (pp. 118–120) in Kevin Ward, Ron Johnston, Keith Richards, Matthew Gandy, Zbigniew Taylor, Anssi Paasi, Roddy Fox, Margarita Serje, Henry Wai-chung Yeung, Trevor Barnes, Alison Blunt, and Linda McDowell, "The Future of Research Monographs: An International Set of Perspectives," *Progress in Human Geography* 33, no. 1 (2009): 101–126.
2 It should not be a contentious issue, but sadly, it often is. Frankly, departmental handbooks should specify as clearly as possible the definition of a scholarly research monograph.

https://doi.org/10.1515/9783111189727-007

book. I will say, though, that if you're ever in the position where publishing a book is necessary for a job, tenure, or promotion, you should get a firm (and written) definition from the department chair and other colleagues stating clearly what will or will not count. The purpose of this chapter *is* to offer some advice on writing research monographs.[3]

Geography has a somewhat checkered history with books.[4] In 1989 John Jakle lamented that geography "has a long tradition of rejecting books as inappropriate vehicles of communication, especially for reporting research."[5] This perception hasn't changed much. In 2011, Ugo Rossi and Barney Warf remarked that "scholarly books (research monographs and essays) appear to be marginalized from the academic landscape, being regarded as 'out-dated', 'time-consuming', and less profitable outlets in terms of career prospects compared to the highly rewarding publications in refereed journals."[6] David Harvey, for example, worried about the lack of "substantive monographs" in the discipline, as did Nigel Thrift, who identified "a general decline in the production of learned books and monographs in favor of journal articles."[7]

Although the comments of both Harvey and Thrift were directed toward the status of geography in the United Kingdom, primarily in response to the paradigmatic Research Assessment Exercise, I believe they speak to geography more broadly.[8] When I first started my academic career, I was offered well-meaning ad-

---

3 In this chapter I focus exclusively on the writing of scholarly monographs based on primary research; I do not address text-books, chiefly because my target audience—graduate students and early-career scholars—is probably not starting off their career writing a text-book.

4 John A. Jakle, "The Writing of Scholarly Books in Geography," in *On Becoming a Professional Geographer*, edited by Martin S. Kenzer (Caldwell, NJ: The Blackburn Press, 2000 [1989]), 124–134; Alison Blunt and Catherine Souch (eds), *Publishing in Geography: A Guide for New Researchers* (London: Royal Geographical Society, 2008); and Ugo Rossi and Barney Warf, "Research Monographs and the Making of a Postdisciplinary Geography," *Dialogues in Human Geography* 1, no. 1 (2011): 103–104.

5 Jakle, "The Writing of Scholarly Books," 124. Jakle's geographic focus in his discussion was directed toward North American geography departments; his comments, however, were (and are) pretty typical for geography programs outside of North America.

6 Rossi and Warf, "Research Monographs," 103.

7 David Harvey, "Editorial: The Geographies of Critical Geography," *Transactions of the Institute of British Geographers* 31 (2006): 409–412; at 410; Nigel Thrift, "The Future of Geography," *Geoforum* 33 (2002): 291–298; at 295.

8 Rossi and Warf, "Research Monographs," 103. For more on the Research Assessment Exercise and its impact on geography, see Noel Castree, "Research Assessment and the Production of Geographical Knowledge," *Progress in Human Geography* 30, no. 6 (2006): 747–782; Keith Richards, Mike Batty, Kevin Edwards, Allan Findlay, Giles Foody, Lynne Frostick, Kelvyn Jones, Roger Lee, David Livingstone, and Terry Marsden, "The Nature of Publishing and Assessment in Geography

vice to *not* write books. Concentrate on publishing refereed articles, I was told. Indeed, my first faculty handbook at Kent State was silent on how to assess books for tenure or promotion. Criteria for tenure and promotion detailed in precise language the quality and quantity of refereed articles required to advance in my profession but it left unclear how a book should count. When the subject arose in faculty meetings, myriad numbers were bandied about. Maybe a scholarly book should count as five articles, someone suggested. But what if chapters had been previously published already, in whole or in part, as journal articles? Wasn't this a deceptive way to double-count one's scholarship?[9]

At this point, perhaps we should ask a more basic question. Why write scholarly monographs?[10] Geography does not have a long tradition of writing books, certainly nothing comparable to history, English, or other humanities-based disciplines. Maybe the lack of "substantive monographs," as Harvey observed, isn't a problem. We always have our refereed journal articles, right?

Alison Blunt presents a positive view about the current and future place of books in geography. First, Blunt writes, "books are important in geography and other disciplines because some research lends itself to a book-length study rather than to journal articles. Books offer the scale and scope to develop an argument and draw on a wider range of material on larger and more detailed scales."[11] This is certainly a major factor in my research and writing. When writing journal articles and book chapters for edited books, I often feel constrained. There's too much I want to say, but I'm limited by pre-determined word counts. However, freed from the constraints of 5,000 or 8,000 words, books offer the opportunity to fully engage with the subject material. A second, more pragmatic reason identified by Blunt is that books "have the potential to shape academic debate."[12] Matthew Gandy concurs, suggesting that the publishing of books by geographers "is critical for the interdisciplinary and international standing" of our discipline. He

---

and Environmental Studies: Evidence from the Research Assessment Exercise 2008," *Area* 41, no. 3 (2009): 231–243; and Tim Hall, "Making Their Own Futures? Research Change and Diversity Amongst Contemporary British Human Geographers," *The Geographical Journal* 180, no. 1 (2014): 39–51.

9 For interested readers, we no longer have strict guidelines on the quantity of refereed articles necessary for tenure or promotion; nor do we (yet) have workable guidelines for published books. The pace of change in departmental procedures is glacial.

10 For other perspectives on the merits of books, see Kevin Ward and Jo Bullard, "Publishing Books," in *Publishing in Geography: A Guide for New Researchers*, edited by Alison Blunt and Catherine Souch (London: Royal Geographical Society, 2008), 28–38.

11 Alison Blunt, "Books and Individual Publication Strategies," pp. 120–121, in Ward et al., "The Future of Research Monographs," at 120.

12 Blunt, "Books and Individual Publication Strategies," 120.

elaborates that "an emphasis on book publishing may begin to redress geography's legacy of marginalization and help to connect the discipline with new audiences both within and outside the academy."[13] Why this is so is because books can punch above their weight. Books allow for more substantial synthesis, reflection, and expansion of ideas that may (or may not) have been published as articles.[14] So far so good. But there's something else—something more visceral that isn't readily experienced when writing other types of academic pieces. Linda McDowell captures it best: "Of all the pleasures in the academic life," she writes, "putting the last few words to the final chapter of a book is perhaps only outweighed by the look and feel of the book when it is finally published."[15]

What a Book is Not

Before jumping down the rabbit hole of writing a book, let's address some misconceptions. First, a *successful* scholarly monograph should not be simply a compilation of previously published manuscripts. Yes, you can use already-published manuscripts as the core of your book. However, the conversion of three to five refereed articles into a book is not a straightforward process.[16] The reason is simple. Journal articles are written as discrete entities, capable of standing alone.[17] Remember, when you write a manuscript such as a journal article it will reflect your current thinking on the subject; in addition, you will make decisions with a specific targeted audience in mind. In turn, these decisions affect the composition, substance, style, and tone of the manuscript. Considerable editing and revising is required to convert previously published manuscripts into a coherent book. Second, a book is not an article on steroids. You just can't take an 8,000-

---

13 Matthew Gandy, "Books, Geography and Disciplinary Status—An Anglo-American View," pp. 107–109, in Ward et al., "The Future of Research Monographs," at 108. I'd be lying if I said I wasn't concerned about the impact my books might have on the discipline or beyond—I'll leave it to the reader to critique my contributions.

14 Linda McDowell, "Why Write Books?", pp. 121–123, in Ward et al., "The Future of Research Monographs," at 123.

15 McDowell, "Why Write Books?" 121–122.

16 There are of course *many* books that are, in fact, simple compilations of previously published essays and articles. Indeed, some 'books' that are cobbled together from previous work do not even attempt to re-write the material or tie the chapters together. Frankly, this isn't really *writing* a scholarly manuscript and I have no more to say about the practice.

17 Louise J. Bracken and Alastair Bonnett, "Research Articles," in *Publishing in Geography: A Guide for New Researchers*, edited by Alison Blunt and Catherine Souch (London: Royal Geographical Society, 2008), 4–12; at 5.

word article and expand the material ten-fold into an 80,000-word monograph. Remember when we talked about balance? In a conventionally structured journal article you might have five main sections: introduction (1,000 words), literature review/theoretical context (1,500 words), methodology (1,000 words), analysis (3,500 words), and conclusions (1,000 words). If we extrapolate to an 80,000-word monograph, we would have to 'flesh out' the sections in the following proportions: introduction (8,000 words), literature review/theoretical context (16,500 words), methodology (8,000 words), analysis (28,000 words), and conclusions (8,000 words). This won't work. And on that point, novice writers need to understand that the process of writing refereed articles and the process of writing scholarly monographs, while comparable in some ways, are not identical. Articles are not books in miniature form and scholarly books are not articles in an expanded form.

It's All About the Structure

To state the obvious: scholarly manuscripts are longer than refereed articles. Much longer. Alarmingly longer. If writing a 5,000-word article is worrisome, writing a 100,000-word book is downright unnerving.

Some of the pressure, I suggest, can be reduced when we think a little bit more on what you might include in a book. I like to tell stories and book-length projects help me tell more in-depth stories. I can introduce more characters (concepts), develop more sub-plots (theses and counter-theses), and include more description (historical and geographical context). In a journal article, I can include only so much of this material, usually in the form of condensed, concise statements supplemented with a string of references. In a book, I can take the time (space) to narrate a more complex and comprehensive story in a way not possible in other formats.

Let's suppose I'm writing about agrarian capitalism. Conceptually, primitive accumulation is a central character. However, even a cursory read of the field will quickly demonstrate that primitive accumulation is a much-contested concept. There is no commonly agreed upon definition and, as a term encountered in several disciplines beyond geography, it has over the years been used and abused in many different ways. How should I develop this character in my work?

By way of illustration, I'll provide two examples, one from a refereed article and another from a book. A few years back I wrote an article for *Gender, Place & Culture*, a feminist geography journal, on "Gender and Sexual Violence, Forced Marriages, and Primitive Accumulation during the Cambodian Genocide, 1975–

1979."[18] Conceptually, primitive accumulation is clearly important—it's in the title! However, other topics and concepts are equally important—if we derive anything from the title: gender and sexual violence, forced marriages, and the Cambodian genocide itself. In terms of composition, I needed to address all these elements, all the while remaining sensitive to competing definitions, debates, and so on. I've indicated already that the concept of primitive accumulation is not without controversy. Suffice it to say, so too the conceptualization of gender/sexual violence and forced marriages is fraught with challenges. Lastly, the article was geared toward geographers, not genocide scholars. It followed, therefore, that I would have to include more background information on the genocide itself than I otherwise would if targeting a journal like *Genocide Studies International*. So, how did I proceed in writing an 8,000-word manuscript?

Primitive accumulation, although a main character, was by necessity presented in abbreviated form, introduced through a series of declarative statements supported by the literature. For example, I write "In general, primitive accumulation refers to the historical process of separating workers from the means of production," followed by key references.[19] A quote by Marx is provided, followed by several other, relatively terse sentences. Essentially, I wanted to demonstrate the centrality of primitive accumulation, keeping in mind, however, that my immediate objective was to address gendered and sexual violence. I just did not have the space to expound upon primitive accumulation as a fully fleshed character. This wasn't the case when I wrote *Red Harvests: Agrarian Capitalism and Genocide in Democratic Kampuchea*.[20] No longer restricted to a limited word count, I could introduce primitive accumulation as a fully developed character. I could provide more detail, both on the conceptual history of the term and on the various transformations of the term in specific historical contexts. I was able to write about the different understandings of primitive accumulation as forwarded by Soviet Marxists, for example, such as Vladimir Smirnov and Nikolai Bukharin; I could elaborate on the differences between so-called primitive socialist accumulation and primitive capitalist accumulation; and I could detail in greater specificity the myriad techniques utilized to separate workers from the means of production. And before you think this background material is nothing more than conceptual fluff, let's see how it fits with my overall purpose in writing *Red Harvests*. I indicate rather quickly, in the preface, that "minimal scholarship has addressed the inner designs

---

18 James A. Tyner, "Gender and Sexual Violence, Forced Marriages, and Primitive Accumulation during the Cambodian Genocide, 1975–1979," *Gender, Place & Culture* 25, no. 9 (2018): 1305–1321.
19 Tyner, "Gender and Sexual Violence," 1308.
20 James A. Tyner, *Red Harvests: Agrarian Capitalism and Genocide in Democratic Kampuchea* (Morgantown, WV: West Virginia University Press, 2021).

of Democratic Kampuchea's [as Cambodia was renamed] agrarian political economy in a theoretically informed, empirically documented study." I explain that this omission is important, "in that most interpretations of the Cambodian genocide fail to articulate with any precision why the Khmer Rouge managed agriculture the way [they] did." In making this argument, moreover, I could detail how most scholars of the Cambodian genocide take as given the 'Marxist' credentials of the Khmer Rouge and thus fail to document with any specificity the conceptual context of what the Khmer Rouge actually did. Hence, I write:

> It is less helpful to try to pigeonhole *a priori* Khmer Rouge ideology and material practice into idealized boxes, that is, to presuppose an essentialist Marxist, Marxist-Leninist, Stalinist, or Maoist doctrine and assess the purity with which CPK [Communist Party of Kampuchea] policies conform or do not conform to these archetypes. Certainly, it is necessary to consider, through empirical documentation, sources of influence on CPK policy and practice, but it is also necessary to position the Marxism evinced by the Khmer Rouge as distinctive. In other words, we should not frame the Khmer Rouge as Cambodian Stalinists or Maoists. We must evaluate the CPK on its own terms, but we must proceed with caution. Fundamentally, this approach requires a robust engagement with the theoretical substance and history of Marxist thought.[21]

The expanded scale and scope of scholarly monographs is both a blessing and a curse. On the one hand, as I've just illustrated, the capacious nature of books allows you to develop more fully several concepts, themes, debates, and so on. On the other hand, the intellectual and organizational effort required is immense; there is a need when writing a monograph to sustain your argument over the course of tens of thousands of words.[22] It's frustratingly easy to get lost in the thickets of your story. Characters (concepts) can take a life of their own and pull you down unexpected paths (debates). This isn't necessarily a bad thing. It does require, however, a firmer grip in steering the manuscript's progression from start to finish. Getting lost in a 5,000-word manuscript is an altogether different problem from getting mired in an 80,000-word tome. For this reason, in my experience, structure is all-important.

Journal articles may, and often do, fall into some relatively conventional formats. Scholarly books rarely do. Writers can, of course, structure their monographs in fairly standard ways, with a clearly articulated introductory chapter followed by chapters on relevant literature, concepts, and case studies. Writers can also include a chapter on methodologies or, alternatively, include this material in an appendix. Conversely, writers can organize their books thematically, historically, or by geographic region. Chapters can be written as self-contained, standalone components or structured as stepping stones leading to a final 'ah-ha' moment in the concluding chapter.

But enough with abstract statements. Let's look at some actual examples. Here's the chapter outline for Simon Springer's book *Violent Neoliberalism: Development, Discourse, and Dispossession in Cambodia:*

Introduction
Part I Development
Chapter One: Violent Politics
Chapter Two: Violent Kleptocracy
Part II Discourse
Chapter Three: Violent Orientalism
Chapter Four: Violent Symbolism
Part II Dispossession
Chapter Five: Violent Accumulation
Chapter Six: Violent Evictions
Conclusion[23]

There is a remarkable symmetry in Springer's monograph. Bracketed by an introduction and conclusion, the heart of the book is organized into three sections (Development, Discourse, and Dispossession), with each section made up of two chapters. These section headings are mirrored in the book's sub-title and form the basis of both the opening and concluding chapters. Indeed, we see development, discourse, and dispossession as leading characters, all of which Springer brings to light early in the story. Springer writes, "The overarching objective is to overcome the divide that exists between the *discursive* legitimation of neoliberalism as a positive element in the *development* of societies and the stark reality of the damage wrought by the adoption of neoliberal politics and the accumulation by *dispossession* that comes attendant to such political economic reform."[24] In one sentence, Springer effectively ties together the main concepts. Subsequently, throughout the remainder of the introductory chapter (and the book itself), Springer always keeps these concepts front-and-center.

---

23 Simon Springer, *Violent Neoliberalism: Development, Discourse, and Dispossession in Cambodia* (New York: Palgrave Macmillan, 2015).
24 Springer, *Violent Neoliberalism*, 16. Emphasis added.

My second example comes from Erik Swyngedouw's monograph, *Liquid Power: Water and Contested Modernities in Spain, 1898–2010*.[25] Based on the title, we might expect his book to be structured chronologically, that is, an historical account of water infrastructure in Spain, spanning the long twentieth century. Here's how Swyngedouw organized his story:

1. "Not a Drop of Water ...": Spain's Cyborg Water World
2. The Hydro-Social Cycle and the Making of Cyborg Worlds
3. *"Regeneracionism"* and the Emergence of Hydraulic Modernization, 1898–1930
4. Chronicle of a Death Foretold: The Failure of Early Twentieth-Century Hydraulic Modernization
5. Paco El Rana's Wet Dream for Spain
6. Welcome Mr. Marshall!
7. Marching Forward to the Past: From Hydro-Deadlock to Water and Modernity Reimagined
8. Mobilizing the Seas: Reassembling Hydro-Modernities
9. Politicizing Water, Politicizing Natures, Or ... "Water Does Not Exist!"

There is a semblance of historical organization but, on closer inspection, we see a more complex structure. With the first sentence in the first chapter, Swyngedouw broadcasts his main objective: "This book explores how water becomes enrolled in the tumultuous process of modernization and development, and how the qualities and powers of water fuse with social, political, and economic processes in the pursuit of social dreams and fantasies nurtured by a diverse set of social actors."[26] From there, Swyngedouw does, loosely, arrange chapters historically; however, specific concepts are developed in the individual chapters, culminating with Chapter 9, whereby Swyngedouw "explores how the preceding arguments open up new avenues from thinking and acting in the context of the escalating water problems and conflicts that plague not only Spain but many other parts of the world as well."[27] In other words, the substantive chapters constitute many trees, which Swyngedouw examines in great detail; the book finishes with Swyngedouw stepping back to gaze upon his conceptual forest, so to speak, and restates his main argument, introduced in the first chapter.

25 Erik Swyngedouw, *Liquid Power: Water and Contested Modernities in Spain, 1898–2010* (Cambridge, MA: The MIT Press, 2015).
26 Swyngedouw, *Liquid Power*, 1.
27 Swyngedouw, *Liquid Power*, 18.

Setting the Stage: Openings

In my discussion on refereed articles, I called attention to the importance of balance. A typical 5,000-word article in *The Professional Geographer*, for example, might include five main sections: Introduction, Theory, Method, Findings, and Conclusions. In this format, the introductory section might be only 500 – 750 words. In a scholarly monograph, prefaces or introductory chapters might run upwards of 5,000 words—the same word count as many refereed articles. Obviously, when writing an introductory section in a monograph as opposed to an introductory section in an article, you can include and expand upon considerably more material. However, I want to highlight another opportunity provided by the monograph, chiefly the opportunity to present material with a bit more flourish or panache. This isn't to deny that writers can be creative in shorter articles. It is to say, though, that when writing monographs, you have more room to include more elaborate openings. Jessica Barnes' *Cultivating the Nile: The Everyday Politics of Water in Egypt* offers a really good example.[28]

A political ecologist, Barnes writes "about the political dynamics that emerge in, through, and around" the Nile River. Her "central argument is that Egypt's water is not a given object of management but rather is made as a resource through daily practices."[29] In clear, concise writing, Barnes positions her study within the fields of anthropology, geography, and other disciplines, and foreshadows her work as important for broader understandings of "society–environment interactions within environmental anthropology, political ecology, and science technologies."[30] But notice what she does next—and why I think this preface is exemplary. Barnes writes, "The book's narrative follows the flow of water through different sites."[31] With this simple sentence, Barnes expressly organizes her narrative to align with the movement of the Nile. She explains, "In Chapter 1, I situate the reader in Egypt, at the end of the river, laying out the key actors who use, manage, and manipulate Egypt's water. ... Chapter 2 focuses on the places of scarcity, where irrigation canals lie empty. ... In Chapter 3, I examine the governance structures that water passes through..." and so on.[32] This technique really shines through when, in the first chapter, she introduces the hydrology of the Nile. She does so in a section titled "The Travels of a Water Droplet." Barnes writes:

---

28 Jessica Barnes, *Cultivating the Nile: The Everyday Politics of Water in Egypt* (Durham, NC: Duke University Press, 2014).
29 Barnes, *Cultivating the Nile*, x.
30 Barnes, *Cultivating the Nile*, xii.
31 Barnes, *Cultivating the Nile*, xii.
32 Barnes, *Cultivating the Nile*, xii.

Rain pours down on a landscape of grassy hills, swampy valleys, and occasional rocky peaks rising to significant heights. It is an August day in the Ethiopian Highlands, midway through the four-month rainy season. A water droplet runs over the saturated soil into the brown, sediment-heavy stream known locally as the Abay, but more widely as the Blue Nile. The river winds a tortuous path as it descends through the highlands, at times cutting a steep-sided canyon into the basaltic plateau. Some 800 km later, after descending more than 1,000 m, the droplet crosses over the Sudanese border. The water slows as it approaches the backed up water behind the Rossieres Dam, 80 km inside Sudan. Through the dam, the droplet continues on its course, flowing through wooded country, at first larger and more varied trees, many of them baobab, giving way to small acacias. Two hundred and fifty kilometers further on, the water meets its second temporary storage point—the reservoir behind the Sennar Dam. The droplet moves through the dam and, some distance downstream, flows into the dusty city of Khartoum. In the middle of this city, at an altitude of 380 m above sea level, the Blue and White Niles meet, continuing their course together. As the river flows northwards, the droplet passes between cultivated fields and pastures of grasses and acacia trees. The surroundings become increasingly arid. Following the Merowe Reservoir and Dam and a series of rapids, the channel widens, marking the beginning of Lake Nasser. The droplet moves through the lake, at some point crossing an invisible border and into Egyptian territory. It reaches the large gates of the Aswan High Dam. Below the dam, the Nile cuts a blue line through a narrow strip of green fields. The droplet flows past temples built thousands of years ago and through the barrages of Esna, Naga Hamadi, and Asyut. The banks become increasingly built up as the river enters the noisy metropolis of Cairo. North of the city, the water passes through the Delta Barrages and the river splits in two, forming the Rosetta and Damietta branches. If the droplet reaches the end of one of these branches, it joins the salty waters of the Mediterranean. It is unlikely, however, to get that far. More probably, at some point, before it reaches the sea, the droplet will be diverted into a canal and onto a field.[33]

It's all there: detailed information on hydrology, geomorphology, geology. In fact, even allusions to geopolitics are made. And it is written in an engaging style that does not detract from but instead enhances the narrative. This isn't stale or boring writing. Neither is it watered down (pun intended) for the reader. One takes away a more intimate knowledge of the region's hydrology and the importance of the Nile. At almost 400 words, however, this stylistic flourish is not always possible in shorter formats.

Barnes' opening effectively sets the scene of her work. However, other openings are equally effective, albeit written from vastly different vantage points. For example, it may be desirable to describe one's research process and to explain how the book materialized. Notable in this regard is Jessica Dempsey's *Enterpris-*

---

33 Barnes, *Cultivating the Nile*, 4–5.

ing Nature: Economics, Markets, and Finance in Global Biodiversity Politics.[34] Avoiding the conventional 'Introduction', Dempsey opens with a section titled "In the Beginning, There Was Failure." She writes:

> In book- and paper-stuffed academic offices, walking down cold and dark streets in Norway alongside government bureaucrats, on Skype interviews with bankers—everywhere I went in the course of my research people talked about the failures of biodiversity conservation. "We tried to make people care about nature for its own sake," said global experts, "without the results." I read about failure within the pages of *Science* and *Nature*; I decoded profound disappointment in the stilted text of multilateral policy documents. Over a beer in a noisy Palo Alto bar, the chief scientist of The Nature Conservancy, Peter Kareiva, explained the problem in his straight-shooting manner, "No one cares about biodiversity outside of the Birkenstock crowd." Biodiversity, he went on to say, "is something that suburban white kids care about and nobody else."[35]

In this brief, opening paragraph Dempsey details how her experiences as a scholar contributed to her work. She frames *a* problem, that of biodiversity conservation, within the pages of scholarly journals and international agencies; and she hints at a personal connection to the problem. She's in touch with key experts. We don't yet know if the conservation of biodiversity is the *only* problem; we're pretty confident, however, that it's probably the main problem. As a scholarly monograph, however, we can expect a number of sub-plots to make an appearance. Dempsey doesn't fail us.

Following a brief digression on the scale and scope of biodiversity loss, Dempsey forwards a crucial question: "How can beleaguered environmental activists, bureaucrats, and ecologists generate the political will to spur governments, business, and the general population to take the urgent action that's needed?"[36] Thus, picking up on the sense of resignation effectively captured in the opening paragraph, Dempsey asserts that "For many ecologists and their allies, the answer lies in a turn toward economics."[37] Ahah! Biodiversity conservation is important— but it is really how ecologists and other experts *promote* conservation that is under examination. More precisely, it's the possible risks attendant to economic approaches to biodiversity conservation that's at issue. To that end, Dempsey articulates the purpose of her work:

---

34 Jessica Dempsey, *Enterprising Nature: Economic, Markets, and Finance in Global Biodiversity Politics* (New York: John Wiley & Sons, 2016).

35 Dempsey, *Enterprising Nature*, 1.

36 Dempsey, *Enterprising Nature*, 2.

37 Dempsey, *Enterprising Nature*, 2.

In this book, I explore this turn to economics, the efforts to speak a new language in global biodiversity conservation. *Enterprising Nature* is a critical exploration of the ascent of what is becoming a new maxim in this field: "In order to make live, one must make economic." In other words, for diverse nonhumans to persist, biodiversity conservation must become an economically rational policy trajectory, sometimes even profitable.[38]

And from there, Dempsey is off and running.

Both Barnes and Dempsey coax readers into their respective monographs. In both instances, a cast of characters is introduced, the scene is set, and dominant plots are identified. What more can one ask of an introduction? Here, by way of conclusion, is a sampling of other thesis statements that serve similar functions:

> Baldly, this book's main argument is that the developing knowledge economy changed and interacted in complex ways with Austin's environment and its system of racial relations: the impact of these transformations has been strongly influenced by a historically varying, but still relatively stable, system of asymmetrical power relations that has engendered both the uneven development among neighborhoods and relative inequalities among peoples.[39]

> This book is primarily concerned with the coevolutionary dynamic of states and firms in shaping industrial transformation, and the role of domestic and international processes in mediating these effects. I go beyond a state-centric view of industrial transformation to consider the interaction between domestic states and national firms in a world economy characterized by deep global economic integration. My aim is to show how the previously one-sided relationship between developmental states and their national firms has changed over time in favor of firms that work with extranational actors embedded in global production networks.[40]

> This book considers the making of ideas of human connection and distance, sameness and difference, belonging and origins in the entangled science, culture, and commerce of human population genetics as they are shaped by and implicated in the politics of identity, belonging, and difference.[41]

> *On Schmitt and Space* is the first English language monograph specifically devoted to studying Schmitt's thought from a geographic perspective. Until now the majority of writings addressing the conceptualization of space in Schmitt has focused on specific aspects of his work ... without attempting to provide an in-depth account of the role of space in his oeuvre as a whole, and of its significance for geographical thought more widely. *On Schmitt and Space* therefore aims to address what we perceive as a substantial gap in the secondary literature,

---

38 Dempsey, *Enterprising Nature*, 3.

39 Eliot M. Tretter, *Shadows of a Sunbelt City: The Environment, Racism, and the Knowledge Economy in Austin* (Athens, GA: University of Georgia Press, 2016), 5.

40 Henry Wai-chung Yeung, *Strategic Coupling: East Asian Industrial Transformation in the New Global Economy* (Ithaca, NY: Cornell University Press, 2016), 3.

41 Catherine Nash, *Genetic Geographies: The Trouble with Ancestry* (Minneapolis, MN: University of Minnesota Press, 2015), 2.

both as a whole and, particularly, in geography, by providing a full investigation of the role of space in his thought and locating it in relation to contemporary debates on the spatial dimension of politics within human geography and related disciplines.[42]

Devoid of context, these sentences in-and-of-themselves fail to capture the complexities of their respective monographs; as heuristic devices, however, these lines illustrate the diverse ways in which authors bring clarity to their monographs in an attempt to keep readers informed of the overall project. For example, Claudio Minca and Rory Rowan's *On Schmitt and Space* effectively positions their project within geography in general, human geography specifically, and its overall contribution to the theorization of space. This focus remains throughout the book; and in fact they open their concluding chapter with the sentence, "The main contention of this book is that spatial concepts have played a central structural role in Carl Schmitt's work."[43]

These examples are direct and to the point. Other monographs open with intimate portraits or vignettes, often based on personal experience. In his *Rule by Aesthetics: World-Class City Making in Delhi*, D. Asher Ghertner opens with three scenes from Delhi in an effort to position his broader study.[44] Scene 2 begins:

> On a hot May afternoon early in my fieldwork in Delhi, I rode an air-conditioned elevator up the Vikas Minar to the office of A.K. Jain, the commissioner of planning in the DDA. Perched inside Delhi's tallest building, whose name literally means 'development tower,' Jain has a nearly panoptic view of the city he governs, and at the time of our meeting Jain was in the midst of guiding his agency in redrafting a document with an equally panoptic vision: the Delhi Master Plan. A statutory document that provides comprehensive planning guidelines for the city, the Delhi Master Plan is updated every twenty years to meet new mandates. According to planning law, Jain and his staff of 26,000 engineers, clerks, and field workers are required to conduct a comprehensive civic survey before any master plan revision can begin. Excited to have secured a meeting with Delhi's top planning official, I opened my notebook to the first of a long list of questions I had prepared on planning practice and began with: How did you survey the city?[45]

Ghertner writes: "Jain's answer that day fundamentally challenged my understanding of planning in India."[46] Similarly, the other two scenes, in his words, "introduce

---

42 Claudio Minca and Rory Rowan, *On Schmitt and Space: Interventions* (New York: Routledge, 2016), 4.

43 Minca and Rowan, *On Schmitt and Space*, 245.

44 D. Asher Ghertner, *Rule by Aesthetics: World-Cass City Making in Delhi* (Oxford: Oxford University Press, 2015).

45 Ghertner, *Rule by Aesthetics*, 3.

46 Ghertner, *Rule by Aesthetics*, 3.

key components of what I call 'rule by aesthetics,' a mode of governing space on the basis of codes of appearance rather than through the calculative instruments of maps, census, and survey."[47] Ghertner's use of scenes is effective for a number of reasons. Notably, the scenes are engaging and draw readers into the story but also into the actual research process itself. One gains a sense of the fluidity of 'being in the field' and of how unexpected comments or encounters may force one to rethink one's approach or guiding assumptions. In addition, the scenes illustrate well the conceptual framing Ghertner develops, this being rule by aesthetics.

Noel Castree, in his monograph *Nature*, also makes use of vignettes, but in a different way.[48] In Chapter 1, titled "Strange Natures," Castree presents seven vignettes that include topics on a startling array of seemingly disparate subjects, such as in-vitro fertilization, rainforests, and Patagonian toothfish. For example, Castree begins one vignette—"Do Fish Have Rights"—with a story:

> In spring 2001 the Texan angling community became the butt of a morally loaded joke. The organization PETA (People for the Ethical Treatment of Animals) threatened to dose a fresh-water fishing retreat, Lake Palestine, with tranquillizer. Why did it do so? In order to put the lake's fish to sleep so that they would not be caught during the Red Man Cowboy Sporting Division Angling Tournament! The tournament had been scheduled for April Fool's Day and a cadre of park rangers were deployed to prevent PETA seeding the lake with sleeping pills.[49]

Castree continues with the story before stepping back and asking if fish should have rights. Collectively, the seven vignettes provided by Castree allow him to speak both to the substantive issue of nature (as a concept) and of how we conceive of nature. Castree writes:

> These seven stories about nature are interesting and memorable. If most of them seem unusual, it's only because most of us rarely pause to consider how deeply insinuated into our thought and practice nature is. The stories above recount only a few of the countless ways in which nature is made manifest in numerous walks of life worldwide. But what do these vignettes actually tell us about the subject of this book? In other words, what lessons about nature can we draw from our seven rather different stories?[50]

And, as one would hope, Castree proceeds to detail the relevance of these stories for his overall project, namely to question our preconceived notions of nature. Swyngedouw's *Liquid Power* offers a variation on this technique. Unlike Castree,

---

47 Ghertner, *Rule by Aesthetics*, 4.
48 Noel Castree, *Nature* (New York: Routledge, 2005).
49 Castree, *Nature*, 4.
50 Castree, *Nature*, 7.

Swyngedouw does not introduce his vignettes until well into the introductory chapter. Five pages into the monograph, Swyngedouw writes, "The stories narrated in this book aspire to move beyond H_2O, the stuff that comes out of taps and irrigation spigots. Instead, the limelight will be on the often-invisible actors and agents that are assembled in and through the flows of water and their benign or erratic behavior."[51] Only then does Swyngedouw narrate "A Few Watery Vignettes" in order to "signal the importance of and active mobilizations of water."[52] Unlike Ghertner or Castree, Swyngedouw uses vignettes in a more circumscribed fashion; the stories are historically and geographically specific, confined to the modernization story Swyngedouw narrates. He does not appear personally as researcher in these narratives; nor do these narratives address water issues beyond the territorial boundaries of Spain. That said, Swyngedouw does remind readers that "Rivers, seas, and aquifers, and the way they are enrolled and transformed, cast wide nets that link together a range of people in very different places, and express often contradictory, conflicting, and intensely controversial processes."[53] In other words, Swyngedouw is ever mindful of casting his conceptual net in waters far from the shores of Spain.

A final note on openings. Writers, myself included, often include a statement as to what the book is not. The hope—my hope—is that readers are less likely to 'mis-read' the monograph and erroneously presume the topic is something other than that intended by the writer. In my monograph *The Geography of Malcolm X: Black Radicalism and the Remaking of American Space*, for example, I wanted to be clear that I was not writing a biography of Malcolm X.[54] Indeed, the first sentences literally are: "This is not a biography of Malcolm X, but rather a geography of knowledge. This is an attempt to place the political thought of Malcolm X within a broader context of fundamental concepts of Geography, including segregation, territoriality, representations of place, scalar politics, and representations of selfhood."[55] In *Wildlife in the Anthropocene: Conservation after Nature*, Jamie Lorimer also provides a caveat as to what his book is not.[56] Five pages into the introductory material, Lorimer writes, "This book is not a synoptic survey of contemporary environmentalism, nor will it have much to say about the 'geo' and the wider range

---

51 Swyngedouw, *Liquid Power*, 5.
52 Swyngedouw, *Liquid Power*, 6–7.
53 Swyngedouw, *Liquid Power*, 7.
54 James A. Tyner, *The Geography of Malcolm X: Black Radicalism and the Remaking of American Space* (New York: Routledge, 2006).
55 Tyner, *The Geography of Malcolm X*, 1.
56 Jamie Lorimer, *Wildlife in the Anthropocene: Conservation after Nature* (Minneapolis, MN: University of Minnesota Press, 2015).

of 'planetary boundaries' threatened by the Anthropocene."[57] Helpfully, Lorimer provides an end-note with readings that do address these topics.

Dramatis Personae; or, Introducing Key Concepts

Often, plays and novels include a *dramatis personae*, or list of characters, in the front-matter. This helps to familiarize the many characters of the story and, to a degree, understand their different roles and functions in the story. Scholarly monographs often have similar sections, although often phrased as "Conceptual Framework" or some related title. Most often, this material appears in the preface or introductory chapter. In *Made in the Philippines: Gendered Discourses and the Making of Migrants*, I include a section titled "Theoretical Sign-Posts."[58] I begin:

> The discursive making of migrants and migration and, subsequently, the construction of policy, should not be viewed as a theoretically abstract exercise, but rather as a guide toward an understanding of power and its material effect on the lives of people. Consequently, a feminist-inspired Foucauldian approach provides a useful starting point. In the remainder of this chapter I highlight four concepts—discourse, institutions, discipline, and subjectivities—that buttress my argument and which will be interwoven throughout succeeding chapters.[59]

Subsequently, throughout the remainder of the introduction, over the span of eight pages, I articulate these four concepts, before I provide an overview of the monograph's chapters in a concluding section entitled "Structure of the Book." This, in fact, is a technique I use quite often, as I do, for example, in *The Politics of Lists: Bureaucracy and Genocide under the Khmer Rouge*.[60] A conceptually dense book, over the course of five chapters, I document how the power of the Khmer Rouge was derived in no small part from the material production of information by a cadre of key revolutionaries. In making this argument, I draw upon multiple and interrelated concepts: lists, documents, archives, bureaucracies, conspiracy theories, networks as metaphors, and lethal surveillance. Unlike *Made in the Philippines*, I do not provide a stand-alone section on 'theories' or 'concepts' but instead introduce these concepts in narrative form, with each subsequent concept building on the former. In fact, the most challenging aspect of writing this book

---

57 Lorimer, *Wildlife in the Anthropocene*, 5.
58 James A. Tyner, *Made in the Philippines: Gendered Discourses and the Making of Migrants* (New York: Routledge Curzon, 2004).
59 Tyner, *Made in the Philippines*, 11.
60 James A. Tyner, *The Politics of Lists: Bureaucracy and Genocide under the Khmer Rouge* (Morgantown, WV: West Virginia University Press, 2018).

was the chronological presentation of these concepts. As writers, we are limited somewhat to a linear structure, even though we know and understand that our concepts operate and interact simultaneously. How do we capture this simultaneity in linear form? Therein lies the source of considerable frustration—and enjoyment! I knew that the presentation of the conceptual material up-front was crucial to the narrative arc of the story. Equally important, I knew that the *writing* of the monograph was dependent on this introductory material. I had the relevant puzzle pieces; the challenge was fitting the pieces together to form a coherent whole. And, like most puzzles, the process was both frustrating and entertaining. That's important. Our frustrations don't have to be (writing) blocks.

The Supporting Cast

I'll be honest. I'm not a fan of stand-alone 'literature review' sections or, in the case of scholarly monographs, entire chapters devoted to secondary literature. My antipathy is not because the material is unimportant but chiefly because the presentation of the material is frequently done poorly. Too often, writers take the easy path and provide a series of sentences that assume the form of 'he said' and 'she said'. This 'laundry list' approach to writing literature is, frankly, about as interesting as reading a laundry list. When shopping, the information provided is important but certainly not compelling.

Much like the supporting cast in a film or novel, secondary literature is vitally important. There's a reason the Academy Awards and the Golden Globes honor actors and actresses in a supporting role. These characters provide depth to other characters; they help with plot development and are influential in setting the tone. Secondary literature in scholarly manuscripts serves a similar purpose. When done well, the secondary literature—the 'stuff' of literature reviews—can help situate the main thesis, clarify contested concepts, and adjudicate debates and disputes within and between disciplinary communities. However, supporting characters should not (normally) take center-stage.[61] If you find in writing your manuscript that the secondary literature is delivering extended monologues, something might be wrong.

In presenting secondary literature in a scholarly monograph, I'd encourage you to think how the material supports *your* main thesis. You're developing concepts; use the supporting cast to *support* your understanding of the concepts. If

---

61 For an interesting look at supporting characters, see David Galef, *The Supporting Cast: A Study of Flat and Minor Characters* (University Park, PA: The Pennsylvania State University Press, 1993).

you claim a particular theory or method is contested, provide examples by drawing on the secondary literature. Above all, don't let the secondary literature steal the scene. An excerpt from Sasha Davis' *The Empires' Edge: Militarization, Resistance, and Transcending Hegemony in the Pacific* provides a good example.[62] In his monograph, Davis addresses "the ways people in ... colonized and militarized spaces *within* the Asia-Pacific region define their places—and the effects these imaginings might have." He argues that "what is at stake today in the region is not just who gets hegemony or control over the vital system. ... Rather, what is being contested is whether hegemony and domination, as practices, can be *transcended.*"[63] In these brief sentences, Davis introduces several concepts—main characters—including hegemony, domination, militarism, and even transcendence. Davis does not interrupt his narrative with a separate 'literature review' section on these various concepts but instead interweaves the various approaches and definitions within his overarching story. A few pages later, for example, Davis presents a section titled "Hegemony-Seeking Power." He begins:

> Hegemony-seeking power is based on exclusion and domination (i.e., inside vs. outside groups, citizens vs. aliens, colonizer vs. colonized, privileged vs. disadvantaged) and therefore has its own particular forms of governance, techniques, economies, politics, and geographies (see, for instance, Foucault, 2007).[64] One of the fundamental strategies of the hegemony-seeking power that is in operation across the Asia-Pacific region is the exclusion of peoples from the realm of the 'citizen' of the colonizing power. ... The scholarship of Giorgio Agamben (1998, 2005) is useful for understanding this colonial situation. In his work, Agamben points out that there is a big difference in the way governments and other powerful entities treat people who are merely physically alive and people who are considered part of the body politic. ... Given the U.S. military adventurism over the past decade in places such as Iraq, Afghanistan, the Horn of Africa, and the southern Philippines ... there are obvious reasons why this theoretical perspective has become popular in contemporary social science research. In this era of the 'War on Terror' and the Obama administration's targeted drone assassinations, Agamben's theories on the marginalization of people and places to the level of disposability has attracted a lot of attention from scholars concerned with geopolitics, warfare, and colonialism (Aguon, 2005, 2006; Coleman & Grove, 2009; Fluri, 2012; Gregory, 2004; Gregory & Pred, 2007; Hannah, 2006; Mbembé, 2003; Minca, 2006, 2007; Sylvester, 2006). ... Thomas Lemke (2005), expounding on Agamben, notes...[65]

Over the space of two pages, Davis introduces the concept of hegemony and, with reference to Foucault, underscores that it is a multi-faceted concept. Davis propos-

---

62 Sasha Davis, *The Empires' Edge: Militarization, Resistance, and Transcending Hegemony in the Pacific* (Athens, GA: University of Georgia Press, 2015).

63 Davis, *The Empires' Edge*, 10.

64 Note: I do not include Davis' references here, as these are not relevant.

65 Davis, *The Empire's Edge*, 17–18.

es to frame his conceptual understanding by drawing on the work of Agamben. This is appropriate, Davis writes, because of how Agamben's work calls attention to several important political, economic, and social conditions currently found in the geographic region. In addition, Davis highlights similar work carried out using Agamben's theory, including an important extension of this scholarship by Lemke. Throughout this and other passages, Davis effectively uses the secondary literature *in support of* his own thesis. We see clearly how Davis' monograph is placed within the wider scholarly community and why his particular conceptual approach is appropriate.

Literary tradition, Kristen Starkowski explains, "treats minor characters as elements of settings, as figures who help us realize major characters more fully through interaction, or as representatives of main themes."[66] I encourage novice writers to approach their secondary literature in a similar way.

The Geographic Focus of Geographic Monographs

Not surprisingly, books by geographers frequently center on particular places. These may be cities, such as Rashad Shabazz's *Spatializing Blackness: Architectures of Confinement and Black Masculinity in Chicago*, or broader regions, such as Davis' *The Empires' Edge*.[67] Regardless, scholarly monographs in geography almost invariably attempt to speak to conceptual issues beyond the immediate geographic focus, that is, the empirically grounded case study. Shabazz, for example, opens *Spatializing Blackness* with a deceptively simple declarative sentence: "This book is about the relationship between people and place."[68] In other words, it is a book about Chicago—but it is decidedly more; and Shabazz wants the reader to understand from the outset that the audience he has in mind is not composed exclusively of readers interested in Chicago. Later in the introductory material, Shabazz writes, "I want to know: What are the consequences that stem from the 'steep rise of these mechanisms' [of carceral power] in the lives of everyday residents? How did this come to be? And why were Black people subject to it? This book is my attempt to answer these questions. To do this I look at Chicago's South Side."[69] Crucially—and necessarily—Shabazz proceeds to explain the salience of

---

66 Kristen H. Starkowski, "'Still There': (Dis)engaging with Dickens's Minor Characters," *Novel: A Forum on Fiction* 53, no. 2 (2020): 193–212; at 196.

67 Rashad Shabazz, *Spatializing Blackness: Architectures of Confinement and Black Masculinity in Chicago* (Urbana, IL: University of Illinois Press, 2015); Davis, *The Empires' Edge*.

68 Shabazz, *Spatializing Blackness*, 1.

69 Shabazz, *Spatializing Blackness*, 2.

Chicago and why this provides an important case study from which to draw conclusions of a more theoretical or conceptual basis.

Davis follows a different path. He opens *The Empires' Edge* with a more geographic-oriented statement: "The Asia-Pacific region has long been an important space between competing powers striving for economic and political dominance."[70] Unlike Shabazz, Davis immediately focuses attention on the region in question and not on other broader conceptual questions. Davis continues in the introductory paragraph detailing the significance of the Asia-Pacific, explaining that the region "hosts an astounding array of military bases, combat-training areas, weapons-testing sites, deployed naval vessels, and nuclear arsenals."[71] That said, Davis soon explains, "in an effort to relieve this region of the heavy toll of militarization, this book presents and promotes a perspective that rejects militarization, contemporary colonialism, and the idea that seeking hegemony is an inevitable condition of international politics."[72] Thus, Davis—similar to Shabazz—remains attentive to the wider impact of his work beyond the immediacy of his case study. Indeed, Davis makes clear his attempt "to situate Pacific activism in the global realm of geopolitics and ask questions about its significance."[73] He concludes, "The small, colonized islands of the Pacific have a story to share. It is a story that can facilitate the efforts of researchers, activists, and residents working to redefine security and transcend the trap of pursuing hegemony."[74] At the risk of sounding like a broken record, both Shabazz and Davis follow a practice I always find helpful: to be mindful of the relationship between the forest and the trees. Sometimes we begin by looking at trees, only to step back and see the forest; other times, we see the forest in order to examine individual trees.

Transitions

In the course of an 80,000-word monograph, it is easy to get lost. Accordingly, writers often make use of transitional sentences or sections to keep readers on track. An obvious technique is to provide a 'path forward' or 'summary of chapters' at the end of the introductory chapter. Swyngedouw, for example, writes, "The following seven chapters develop the occasionally tragic, sometimes heroic, but always dramatic unfolding of Spain's modernization through a combined transformation

---

70 Davis, *The Empires' Edge*, 1.
71 Davis, *The Empires' Edge*, 1.
72 Davis, *The Empires' Edge*, 1.
73 Davis, *The Empires' Edge*, 33.
74 Davis, *The Empires' Edge*, 33.

of social and environmental relations."[75] Lorimer also includes a section titled "Structure of the Argument" and begins,

> The first three chapters outline and illustrate the conceptual foundations of the approach to conservation I have summarized. In chapter 1, I present the ontology of wildlife that forms the foundations for this book. I illustrate this with reference to Asian elephants and the political ecology of Sri Lanka. In chapter 2, I explore conservation as a process of learning to be affected and present the concepts of non-human charism and affective logics in conservation.[76]

Transitions are necessary, however, beyond the introductory 'structure' of the book. Indeed, when I revise and rewrite scholarly manuscripts I pay special attention to the transitions from section to section, chapter to chapter. This often entails the use of direct *linkage* statements. In transitioning from the first to second chapters, for example, Swyngedouw concludes the former by writing, "Before we embark on this mission, the next chapter will briefly explore the theoretical and conceptual perspectives that have both inspired and guided our archeology of Spain's hydro-social past and present, and that charts the contours of its future."[77] Note that this statement appears at the end of Chapter 1, with Swyngedouw already informing us that Chapter 2 "presents a range of theoretical arguments and conceptual heuristic tools that have played a central role in weaving the storylines together."[78] Our takeaway is that Swyngedouw is reminding us constantly of the objective of the book—its story-line—and how the various sections and chapters fit together in his narrative. He begins Chapter 2: "Taking Spain's postimperial condition as its entry point, the book seeks to elucidate the relationship between the hydro-social cycle—the socially embedded techno-institutional organization of the material flows of water—and the process of modernization as it unfolded during the twentieth century."[79] This is a powerful opening sentence and deserves some attention. Swyngedouw reminds readers of his overall argument—without parroting material from Chapter 1. He explains that Chapter 2 provides the necessary theoretical, or contextual, material that will shape the chronological account that follows. In addition, Swyngedouw is clear that while Spain is the empirical focal point of the book, he intends to address topics that extend beyond his case study. At this point, attentive readers should have no problem following the course laid out by Swyngedouw.

---

75 Swyngedouw, *Liquid Power*, 14.
76 Lorimer, *Wildlife in the Anthropocene*, 15–16.
77 Swyngedouw, *Liquid Power*, 18.
78 Swyngedouw, *Liquid Power*, 14.
79 Swyngedouw, *Liquid Power*, 19,

Chapter by chapter, Swyngedouw provides concise transitional statements that helpfully guide the reader from beginning to end. In doing so, the reader's focus remains keyed to the main themes and concepts. There is little confusion with regard to Swyngedouw's purpose nor of his path to achieve that objective. To achieve this level of clarity requires considerable foresight in the planning phases of the manuscript and critical reflection in revising subsequent drafts.

Coming to Closure; or, All Good Things Must Come to an End

Scholarly monographs are long, but they don't continue indefinitely. At some point, the author must bring the book to a close. This doesn't preclude the writing of a *second* book, to continue arguments not fully explored in the first. It does mean (usually) that the book exists as a self-contained unit. Consequently, writers often provide a concluding chapter. These may be framed expressly as 'conclusions' or, alternatively, as an 'epilogue.' Alternatively, writers may simply end the monograph, rather bluntly, making no attempt to bring the book to closure.[80] This latter approach, I think, generally makes for a less-than-satisfactory monograph. Having journeyed with the writer for tens of thousands of words, the reader may feel frustrated in not fully knowing how it all ends.

Concluding chapters, in brief, permit the writer to draw together the various conceptual elements that comprise the book. Stated differently, conclusion chapters are *usually* written from the vantage point of seeing the forest from the trees. That said, what is included in the concluding chapter can vary greatly. For some authors, it is important to recount, almost step by step, the journey just taken. The guiding thesis is restated and perhaps a brief recap of the key conceptual grounding of the monograph is provided. Beyond this, the concluding chapters allow the writer to clarify conceptual connections and identify areas of further research. By way of illustration, Ghertner opens his concluding chapter with a simple, declarative statement: "This book began with the paradox of a state encumbered by its own vast calculative apparatus."[81] Then, after describing some of the particularities of Delhi, Ghertner steps back: "*Rule by Aesthetics* has examined

---

80 In *Landscape, Memory, and Post-Violence in Cambodia* (Lanham, MD: Rowman & Littlefield, 2017), I do not provide an end-chapter that brings closure to the conceptual material. I still regret my decision not to include a concluding chapter. Notably, I also do not include a concluding chapter to this book you're reading now. My choice was deliberate. Instead of telling a story on writing, I envisioned this monograph as a tool-kit, with each chapter forming a stand-alone discussion on particular modes of writing.

81 Ghertner, *Rule by Aesthetics*, 183.

the aftermath of Delhi's calculative crisis and the novel ways in which its various governors were able to turn what appeared to be a weakness into a strength."[82] Ghertner continues to recap his main findings and how these fit into his overarching thesis, this being the "rule by aesthetics" that is a "mode of partitioning space on the basis of codes of appearance."[83] Lastly, Ghertner provides an update on events in Delhi and concludes his monograph with a series of questions based on his case study: "Will the world-class city follow the inclusion that elite property owners and developers have long clamored for, wherein the social worth of subjects is linked to their ability to pay for property? Or will the urban poor, now subjects of the discourse of the world-class city, bend its trajectory, forging a more inclusive vision of the city to come?"[84]

By way of contrast, Springer concludes *Violent Neoliberalism* with a theoretical discussion of neoliberalism and makes little direct reference to his empirical study of Cambodia. Springer underscores that "while the contextual specificities of Cambodia's particular experiences with violence, neoliberalization, and democratic processes are undoubtedly a major concern," he argues that "the violent geographies of neoliberalism [have a] relational quality that stretches across multiple sites."[85] Hence, Springer concludes, "Such an acknowledgement brings a wider resonance to my argument, where Cambodia provides a window on the patterns of violence that are associated with the socio-spatial transformations that are occurring in a range of sites undergoing neoliberalizing processes all across the globe."[86] On this point, Springer effectively ties back to his opening sentence of the monograph, precisely that "Neoliberalism has become the dominant political economic arrangement in our world today."[87] Springer begins his book looking at the world and he concludes his book looking at the world.

In bringing this section to a close, I encourage you to pay special attention to the symmetry of beginnings and endings. In other words, it's important to provide balance, in terms of substance, composition, tenor, and style, between your first and last chapters. If, for example, you begin your monograph with a specific (memorable) metaphor, you should try also to finish your monograph by returning to the metaphor. Remember Barnes' journey of a water droplet? Here's part of her concluding chapter:

---

82 Ghertner, *Rule by Aesthetics*, 184.
83 Ghertner, *Rule by Aesthetics*, 184.
84 Ghertner, *Rule by Aesthetics*, 198.
85 Springer, *Violent Neoliberalism*, 168.
86 Springer, *Violent Neoliberalism*, 168.
87 Springer, *Violent Neoliberalism*, 1.

A complex set of social, biophysical, technical, and political processes, operating on a range of scales, mold the course of the river's water. That water can be traced back to a distinct source: rainclouds over the highlands of East Africa. But the way in which the Nile passes through Egypt, making Egypt's portion of the Nile Basin, is integrally tied to a number of technologies, management decisions, and agricultural practices.[88]

The connection is subtle, but we sense the presence of Barnes' humble water droplet. Its journey—and ours—is complete.

Some Final Thoughts on Writing a Scholarly Monograph

Throughout this chapter I have highlighted several elements and techniques that hopefully will aid in the drafting of your monograph. From beginning to end, a high degree of awareness is necessary to aid the reader in their journey. Similar to journal articles and book chapters written for edited volumes, scholarly monographs require considerable care to ensure continuity and cohesion. Unlike these other formats, however, the inclusion of myriad plots and sub-plots, conceptual characters, and a supporting cast, raises the ante in organization and style. And that is really the key to writing scholarly monographs. Writing a book is like sailing open waters. Without adequate navigational aids, it is too easy for a writer to be cast adrift and float sullenly in a sea of emotional frustration.

But let's not dwell on the fear of becoming lost at sea. I enjoy writing scholarly monographs precisely because of the wide-open expanses of uncharted waters. Freed from the narrow channels of journal articles or the rivulets of encyclopedia articles, the writing of scholarly monographs can be invigorating and exhilarating. And so, raise anchor and let the voyage begin!

88 Barnes, *Cultivating the Nile*, 169.

Chapter 6 Book Reviews

Let's be honest. Book reviews, both in geography and beyond, often get a bad reputation. As Franklin Obeng-Odoom bemoans, book reviews are "discouraged because they are not research-based or because they constitute summaries, not anything new." Likewise, book reviews are seemingly easy to write and thus aren't serious pieces of scholarship. For these reasons and more, "book reviews are pushed to the margins of academic activity."[1] Päivi Oinas and Samuli Leppälä are equally blunt in their assessment: book reviews don't seem to be anyone's first priority.[2]

In this chapter I hope to convince you otherwise. Book reviews are vital to academia and we should afford this type of writing the respect it deserves. To fully appreciate my argument, let's first think about why we need book reviews. Only by understanding the importance of *reading* book reviews can we understand better how to *write* book reviews.

Book reviews are of different kinds, and they can serve multiple purposes.[3] Broadly conceived, though, book reviews serve two primary functions. First, a book review offers insight if the book under review can possibly be used in a classroom setting, for example as a key text in a graduate seminar or as a supplement to other texts. Second, a book review informs readers if the book under review is useful for their research. Similar to abstracts, good book reviews provide a form of filtering, whereby scholars can sift through the voluminous scholarship on a topic. There is, however, a third—and under-appreciated—purpose: book reviews are often the terrain where academic debates play out. Let's begin.

The Components of a Book Review

"There is much more to writing a book review," Helen Jarvis explains, "than meets the eye." Indeed, she underscores that "because the length is usually quite short (in the range of 400–1200 words) this piece of writing has to be succinct, not dense, and needs to be critically engaging in a constructive rather than polemic way."[4]

---

1 Franklin Obeng-Odoom, "Why Write Book Reviews?" *Australian Universities' Review* 56, no. 1 (2014): 78–82; at 78.
2 Päivi Oinas and Samuli Leppälä, "Views on Book Reviews," *Regional Studies* 47, no. 10 (2013): 1785–1789.
3 Oinas and Leppälä, "Views on Book Reviews," 1786.
4 Helen Jarvis, "Book Reviews," in *Publishing in Geography: A Guide for New Researchers*, edited by Alison Blunt and Catherine Souch (London: Royal Geographical Society, 2008), 16–18; at 17.

https://doi.org/10.1515/9783111189727-008

And while there is no uniform template or structure for writing a book review, there are certain components that are typically found. If a particular composition is required of a book review, general rules are spelled out in the journal's guidelines.

Book reviews that appear in *Human Geography*, for example, should adhere to the following recommendations:

> Book reviews focus on a single book (though often engage with other literature to address broader questions and themes within geography). The purpose of a regular book review is to summarize the main findings of the book and discuss how it contributes to debates in critical/radical geography (theoretically, methodologically, pedagogically, etc.). Book reviews are a maximum of 2000 words, inclusive of footnotes and references. Book review titles should be formatted as such: Title of Book, Book's Author(s), Place of publication or location of source, Publisher (if applicable), Year (or date if an exhibit).[5]

Here, the guidelines clarify that book reviews focus on a single book but it is desirable that the review situate the book under review within the larger context of geography. Book reviews should not describe the content of the book, that is, merely to re-present the table of contents, but rather to *engage* substantively with the content of the monograph. *Human Geography*, despite its fairly expansive title, is actually a specialized journal, one that explicitly "favors political as well as theoretically-based articles [written] from various left positions that definitely include socialism."[6] Consequently, the guidelines for book reviews published in *Human Geography* suggest that reviewers demonstrate how the reviewed monograph contributes to key debates in critical and/or radical geography. The inclusion of the book's title, author(s), publication date, and so on are boiler plate material, common to all book reviews. Notably, *Human Geography* also publishes book review essays. These differ in that the review should explicitly compare and contrast two or more books. As specified in the guidelines:

> The purpose of book review essays is to discuss emerging conversations in the field. They summarize the main findings of each book, but also explain broader debates in the field under review. Book review essays are a maximum of 5000 words, inclusive of footnotes and references. For review essays, create an appropriate title of no more than 25 words, followed by the sources in the same manner as book reviews.[7]

---

5 *Human Geography*, "Manuscript Submission Guidelines," available at https://journals.sagepub.com/author-instructions/HUG (accessed April 14, 2021).

6 *Human Geography*, "About This Journal," available at https://journals.sagepub.com/home/hug (accessed April 14, 2021).

7 *Human Geography*, "Manuscript Submission Guidelines," available at https://journals.sagepub.com/author-instructions/HUG (accessed April 14, 2021).

A third format is illustrated by the American Association of Geographers' *AAG Review of Books*. Launched in 2013, the *AAG Review of Books* replaced the conventional book review format of the two flagship journals of the AAG, the *Annals of the American Association of Geographers* and *The Professional Geographer*, with three formatting styles: (1) scholarly reviews of current books related to geography and cognate fields; (2) review essays on several books on a single topic; and (3) a book review composed by several commentators, including a response from the monograph's author(s). More precisely, book reviews, regardless of format, should clearly and concisely identify the monograph('s) importance, the prime objectives of the monograph(s), and the extent to which these objectives are realized. Book reviews should not merely report the content of the book but instead provide a critical evaluation of the book's content. Chapter-by-chapter summations are to be avoided. Lastly, in terms of intended readership, the *AAG Review of Books* expects that reviews will be written in such a way as to appeal to a broad cross-section of geographers and scholars working in cognate disciplines. The presumption is that readers are interested, fundamentally, in (1) knowing if the book is worth reading; and (2) understanding how the book contributes to the progress, conception, and stature of geography or relative disciplines.[8]

As this brief overview of two book review forums illustrate, there is no generic template, although book reviews hold in common two key components: a consideration of the book's applicability to teaching and to scholarly research; and a consideration of how the book is situated within broader disciplinary and cross-disciplinary trends and debates. Beyond that, book reviews run the gamut in terms of style and tone. Some reviews are matter-of-fact and provide a straightforward distillation of the book's strengths and weaknesses. Other reviews are more contemplative, both in scale and scope, and offer more reflective evaluations of the reviewed book. Regardless of the approach taken, however, you—as a potential writer of book reviews—should always keep in mind why anyone would want to read a book review. Remember: it isn't so much that scholars want to read *your review*; rather, they want to know if they should read *the author's book*.

The Components of Book Reviews

Thus far, I've explained that book reviews can vary tremendously but, at their core, all typically provide an assessment of the book's use in one's research or teaching

---

8 *AAG Review of Books*, "Guidelines for Reviewers," available at http://www.aag.org/AAGRB_Re viewer_Guidelines (accessed April 14, 2021).

activities and, more broadly, an evaluation of how the book fits into larger scholarly trends and debates. Now, I'll provide some concrete examples and highlight several stylistic techniques. We'll begin, reasonably enough, with ways to start a book review.

Beginnings

In a review of Frank Moulaert and Allen Scott's edited book *Cities, Enterprises and Society on the Eve of the 21st Century*, Richard Smith opens: "This edited collection is based on the papers given at a conference of the same name in Lille (16–18 March 1994)."[9] Simple, direct, and straightforward, Smith informs us that the edited volume is the outcome of a conference. Smith continues: "The book has a clear exposition and wide scope through division into four parts and, as the blurb tells us, provides 'an up-to-date review and restatement of concepts and analytical insights about the relations between the dynamics of the production system and urban society.'"[10] By way of comparison, Don Mitchell begins his review (published in *Human Geography*) of David Harvey's *Rebel Cities: From the Right to the City to the Urban Revolution* as follows:

> "Value—socially necessary labor time—is the capitalist common," David Harvey writes (p. 77) in what is undoubtedly the central argument of this collection, and it is represented by money, the universal equivalent in which the common wealth is measured. The common is not, therefore, something that existed once upon a time and has since been lost, but something that is, like the urban commons, continuously being produced.[11]

Mitchell, here, begins his review not with a declarative statement of Harvey's book. Instead, Mitchell calls attention to an argument made by Harvey well into his monograph (page 77). In this example, Mitchell's lead makes sense, given the orientation of *Human Geography* to deliberately evaluate books from a 'left' perspective and how reviewed books relate to on-going debates in critical or radical geography. However, I'd maintain that Mitchell's opening is effective regardless of the venue, for he immediately draws the reader into an on-going debate. On this point, I'm not downplaying the more conventional and forthright beginning of Smith. I'm simply suggesting, at this point, that novice writers of book reviews consider these almost

---

9 Richard Smith, "Review of *Cities, Enterprises and Society on the Eve of the 21st Century*, edited by Frank Moulaert and Allen J. Scott," *Area* 31, no. 2 (1999): 189–190; at 189.

10 Smith, "Review of *Cities*," 189.

11 Don Mitchell, "Book Review: *Rebel Cities: From the Right to the City to the Urban Revolution*," *Human Geography* 7, no. 1 (2014): 117–120; at 117.

as two ends of a spectrum. Both styles are effective and both convey important information.

Here's some additional opening lines from a scattering of book reviews. In each instance, it's possible to quickly gain a sense of how the reviewers decided to organize their reviews.

> This represents a long-awaited contribution from fluvial geomorphologists within the British Geomorphological Research Group to the series of publications arising from the Geological Conservation Review, initiated by the Nature Conservancy Council in 1977.[12]

> It's good news that Mike Davis has written a new book about Los Angeles![13]

> In most countries water has ceased to be the sole domain of scientists.[14]

> When I recently ran a field trip to Florida, my co-instructor asked me some basic questions about Florida's early history and I realized that I knew almost nothing that happened in my home state prior to the twentieth century.[15]

> I write this review 28 weeks pregnant with my second child, deeply enmeshed in the world of preparing for a hospital birth whilst being an active researcher in the world of food and a self-professed 'foodie.'[16]

The first example, provided by David Sear, quickly places the reviewed edited volume within its genesis. However, Sear indicates also that the publication was expected, thus providing additional information on the book's origin. Dear's jocular opening establishes immediately that the author under review, Mike Davis, has written many books and implies that Davis' prior books have been well received. Dear's beginning, also, lets the reader know that the review will *probably* provide some discussion of Davis' contributions beyond the present book. Unlike Sear and Dear, Rachel McDonnell—in our third example—makes no reference either to the book under review or to its author. Instead, she opens with a provocative assertion that immediately raises questions both of the salience of water and of the contestations surrounding the epistemological 'knowing' of water. Lastly, in my final two

---

12 David Sear, "Review of *Fluvial Geomorphology of Great Britain*, edited by Ken Gregory," *Transactions of the Institute of British Geographers* 24, no. 1 (1999): 125–126; at 125.

13 Michael Dear, "Review of *Set the Night on Fire: L.A. in the Sixties*, by Mike Davis and Jon Wiener," *The AAG Review of Books* 8, no. 4 (2020): 196–198; at 196.

14 Rachel McDonnell, "Review of *Watershed*, by Ticky Fullerton," *Progress in Physical Geography* 27, no. 1 (2003): 143–144; at 143.

15 Jason Dittmer, "Review of *Liquid Landscape: Geography and Settlement at the Edge of Early America*, by Michele Currie Navakas," *Social & Cultural Geography* 19, no. 6 (2018): 831–832; at 831.

16 Miriam Williams, "Review of *A Bun in the Oven: How the Food and Birth Movements Resist Industrialization*, by Barbara Katz Rothman," *Gender, Place & Culture* 24, no. 12 (2017): 1807–1808; at 1807.

examples, Jason Dittmer and Miriam Williams begin with personal reflections, but in very different ways. Dittmer begins with the startling realization that he doesn't really know his home state as much as he thought—a very geographic observation. Williams, conversely, situates her embodied experiences as a scholar, an expectant mother, and a 'foodie' in relation to the reviewed book.

Five different book reviews and five different beginnings: all are effective in their own way.

The Dreaded Chapter Summary

Perhaps the least interesting component of a book review, both to read and to write, is the near-ubiquitous summation of chapters. In fact, it's because this component is often so dull in presentation that some outlets, such as the *AAG Review of Books*, actively discourage reviewers from including chapter summaries. That said, a summation of chapter contents can be vitally important, especially if one is considering using the reviewed book in a classroom setting. Book reviewers, therefore, must make two key decisions. Should a chapter summation be included? And, if so, how should it be presented?

If you've decided not to include a chapter summation, you can probably skip this section. For everyone else, let's see how we can most effectively convey the material without reducing readers to a state of boredom. McDonnell provides an overview that, while not addressing each chapter, does cover the contents in considerable detail. She writes:

> The chapters of the book tackle individual water issues and bring together the often conflicting views of the groups and personalities involved. The second chapter covers the events of the Cryptosporidium scares that faced Sydney from July to September in 1998 disrupting water supply and usage. ... City water distribution and usage remains the topic for the third chapter, where the various sides of the debate on private or public ownership of water management organizations are discussed. Water problems of rural areas become the focus for the next four chapters, with useful descriptions of the usage of the Murray-Darling river basin, salinization, river damming schemes and the politics behind irrigation and water allocation ownership. Politics at all levels—inter-owner, inter-state, state–federal government —are discussed in more detail in the next few chapters.[17]

Notably, McDonnell does not provide an in-depth summation of each chapter. She also doesn't provide a summary statement of each chapter. Indeed, McDonnell doesn't even tell us how many chapters are included in the book! And this, I

---

17 McDonnell, "Review of *Watershed*," 144.

think, is fine. It's not difficult—especially in our digital world—to find the table of contents for books. Instead, McDonnell highlights a mixture of themes, concepts, and geographic locales and, in doing so, efficiently conveys the substance of the monograph.

To see how one might highlight each chapter, we can turn to Tracey Skelton's review of Mike Crang's *Cultural Geography*. Following a brief introduction, Skelton writes:

> The book consists of eleven chapters that focus on some of the key themes within contemporary human geography. The first defines and debates the concept of 'culture' as a subject for study and analysis in its own right and also how it relates to geographical questions. It is a fairly comprehensive chapter, and establishes the difficulty and yet the usefulness of the term. Chapters 2, 3, and 4 focus on the subject of 'landscape'—the ways in which landscapes interact with people and time; the significance of symbolic landscapes and the role of literary landscapes. I feel that Chapter 2 will be particularly useful for students as an introduction to ideas and questions surrounding the study of landscapes. ... The remaining themes described in the book cover the concept of self and 'other' (Chapter 5), media forms and environments (film, TV and music in Chapter 6), and a very good and clear discussion of space and place (Chapter 7). Chapters 8 and 9 introduce geographies of commodities and consumption, and cultures of production respectively. Outlined in Chapter 10 are the complex issues of nationhood, the meanings of homelands and the ways in which continual circular and cross-connections are made by people and cultures as space and place are transcended. The final chapter deliberately avoids being a conclusion to the book itself, but rather considers the ways in which knowledges are constructed.[18]

In this example, Skelton provides a step-by-step summary of individual chapters but, importantly, she interjects her personal opinions ("I feel that ...") and provides a running commentary on the relative merits of each chapter ("Chapter 2 will be particularly useful ..."). In other words, the dreaded chapter summation isn't just a skeletal framework but instead a fully fleshed component that *animates* the review. Williams, in her review of *A Bun in the Oven*, adopts a similar style. Each of the ten chapters under review is singled out and each is accorded its own descriptive statement. Williams writes, for example, "Chapter Two comprehensively documents the history and changing role of the midwife over time." However, she follows each statement with an indication of the chapter's significance in the overall project. Williams writes: "Here Rothman argues for a renewed understanding of midwives as skilled artisans of birth, much like food artisans, who carefully learn a craft through thousands of hours of hands-on-experience."[19]

---

18 Tracey Skelton, "Review of *Cultural Geography*, by Mike Crang," *Area* 31, no. 2 (1999): 190–191; at 190.
19 Williams, "Review of *A Bun in the Oven*," 1807.

Ultimately, unless guidelines specify one way or another, the decision to include a section that summarizes individual chapters is largely one of personal preference. If you, as reviewer, determine that the reviewed book is most appropriate for teaching activities, a more-or-less fully developed summation of chapters may be warranted. On the other hand, if the main contribution of the book lies elsewhere, an abbreviated description of selected chapters is reasonable. Lastly, keep in mind that books and edited books are not limited in the functions they serve; any given book can contribute both to teaching and research. If you determine a book does address both functions, that should be clearly detailed in the review.

Contributions to Teaching Activities

Apart from text-books, readers may not always be sure if a given monograph is appropriate for use in a classroom setting. More so, readers may question if the book is appropriate for, say, undergraduate students, advanced undergraduate students, or graduate students. On this point, book reviewers should always be sensitive both to the appropriate grade level and to the different learning contexts of students.

In terms of presentation, it is common that overall assessments of a book appear *after* the introduction and even *after* the chapter summary section. Ray Bromley adopts this structure in his review of Julie Urbanik's book *Placing Animals: An Introduction to the Geography of Human–Animal Relations.*[20] In the second paragraph, Bromley acknowledges that "Urbanik has written *Placing Animals* as a textbook" and, perhaps for this reason, Bromley devotes considerable space to individual chapters.[21] Subsequently, Bromley suggests that the book "will appeal to students willing to debate ideas and confront a wide range of esoteric examples, drawn from across the world. It raises many moral and ethical questions that should take students outside their comfort zones, questions that are much better explored in small-group discussions or in debates and essays, rather than in large lecture halls and multiple-choice tests."[22] This is a particularly helpful review, for Bromley goes beyond a simple determination that the book is appropriate for undergraduates. Indeed, Bromley suggests also that *"Placing Animals* has many of the

---

20 Ray Bromley, "Review of *Placing Animals: An Introduction to the Geography of Human–Animal Relations,* by Julie Urbanik," *The AAG Review of Books* 2, no. 4 (2014): 133–135.
21 Bromley, "Review of *Placing Animals,*" 133.
22 Bromley, "Review of *Placing Animals,*" 134.

currently fashionable features of textbooks, notably a series of thought-provoking insets to the text: profiles of the tiger, the camel, the dog, the donkey, the pig, and the whale and dolphin. Each chapter ends with lists of Discussion Questions, Keywords/Concepts, Resources, References, and some exercises listed as 'Practicing Animal Geography.'"[23] In short, Bromley provides a thorough review of the book and the reader has much material to consider when deciding if the book is appropriate for their course. Also helpful, Bromley considers how *Placing Animals* might contribute to research activities. Bromley clarifies, "Researchers will find *Placing Animals* thought-provoking in the range of ideas and issues that it covers, but sometimes frustrating because, as a textbook, depth and detail are very limited on many topics."[24]

Contributions to Research Activities

Scholarly monographs are often structured around one or two in-depth case studies. However, because of specific theoretical, conceptual, or methodological contributions, monographs often speak to wider audiences. For example, in much of my work on the Cambodian genocide, on the one hand, I want to contribute to the empirical understanding of the Khmer Rouge and of the political economy of Democratic Kampuchea. On the other hand, I hope to contribute to scholarship on genocide and mass violence more broadly. In other words, research monographs (especially) speak both to 'forests' and 'trees,' and this aspect should be conveyed in book reviews.

Janice Monk provides a review of Pratyusha Basu's *Villages, Women and the Success of Dairy Cooperatives in India: Making Place for Rural Development.*[25] In her review, Monk highlights the contributions of Basu's work and underscores both its conceptual and methodological strengths. Monk writes: "[Basu] advocates the importance of bringing into analyses geographic sensibilities in which the politics and practices of place, space, scale, class, caste, and gendered labor practices and cultures come together to reveal more nuanced histories of the [cooperative] initiatives." Monk underscores that "Fine-grained ethnographic details are the key to [Basu's] analysis, although they do require sustained concentration by the reader to absorb. Tables and figures comparing the spatial and political histories and

23 Bromley, "Review of *Placing Animals*," 134.
24 Bromley, "Review of *Placing Animals*," 134.
25 Janice Monk, "Review of *Villages, Women and the Success of Dairy Cooperatives in India: Making Place for Rural Development,* by Pratyusha Basu," *The Professional Geographer* 64, no. 3 (2012): 464–465.

patterns, political participation, and management practices are helpful in tracking the differences between the villages." Monk continues: "Basu's methodology combines detailed observations and interviews, the latter revealed especially well in extracts of in-depth conversations, which greatly enriches the book."[26] Notably, Monk concludes,

> The book ... has wider implications for development studies beyond the case of cooperative dairying in India. Basu's research is critical of broad-brush studies, policies, and programs of international and national development that aim to alleviate poverty with 'one-size-fits-all' perspectives and that are not sufficiently attentive to diversity within the community cultures and politics or open to multifaceted approaches to sustaining livelihoods.[27]

Overall, Monk provides a thorough review that greatly informs the reader of key theoretical frameworks, concepts, and methods; and articulates why and how the manuscript extends beyond the case study.

Let's be frank. I am probably not going to conduct primary research on dairy cooperatives in India. So why would I reach for Basu's book? The reason is because of the research extensions—and Monk effectively highlights myriad reasons why I should read the book. For example, in my work on the Cambodian genocide, I devote considerable attention to agricultural cooperatives initiated by the Khmer Rouge. The different contexts of cooperative farming in late twentieth-century India and forced labor practices in Democratic Kampuchea during the late 1970s are substantial; and yet, parallels that I otherwise might overlook *without* reading Basu's monograph (or Monk's review) exist. For example, Basu's emphasis on "geographic sensibilities" that articulate the intersections of place, space, scale, class, and gender resonate loudly, not only with my work but, I imagine, with that of other scholars working in development and development-related studies. An effective review, such as Monk's, provides an opportunity for scholars working across disparate fields, in different time periods, and in different geographic regions, to make connections that otherwise might go unremarked. This is an invaluable contribution to the practice of geographic research made possible by a book review.

Alexander Diener's review of Alison Mountz's *Seeking Asylum: Human Smuggling and Bureaucracy at the Border* likewise offers guidance for writing book reviews helpful for scholarly activities.[28] From the opening statement, Diener highlights how Mountz's book can inform scholarship in myriad ways. Diener writes,

---

26 Monk, "Review of *Villages, Women*," 465.
27 Monk, "Review of *Villages, Women*," 465.
28 Alexander C. Diener, "Review of *Seeking Asylum: Human Smuggling and Bureaucracy at the Border*, by Alison Mountz," *The Professional Geographer* 63, no. 4 (2011): 563–565.

"In *Seeking Asylum*, Alison Mountz offers a theoretically informed treatise on the changing landscape of 'irregular migration' that will be of interest to geographers, human rights scholars, and border studies experts alike." Diener continues: "Her deft employment of an ethnographic methodology and application of the geographic lens cultivates new knowledge on a topic requiring fresh ideas." Notice that Diener situates Mountz's work broadly in terms of disciplinary specialties ("geographers, human rights scholars, and border studies experts") *and* in terms of research trends ("a topic requiring fresh ideas").[29]

As a final example, let's turn to Lucy Jarosz's review of Susanne Freidberg's *French Beans and Food Scares: Culture and Commerce in an Anxious Age.*[30] Jarosz begins with a substantive overview of the book, highlighting Freidberg's empirical work on "green bean commodity networks" operating in the global economy. Jarosz identifies the broader extensions of Freidberg's work. Jarosz, though, then pivots to the wider applicability of the monograph. Jarosz writes: "This book makes several important contributions to the growing research literature on food networks that connect production in the Global South to consumption in the North" and, in addition, the book "breaks new ground methodologically by siting ethnography in multiple spaces and places in order to reveal the importance of considering the spatialities of networks and comparing the socioeconomic relations of production and consumption for the same commodity produced and consumed in different places."[31] On this latter point, according to Jarosz, Freidberg

> makes an important contribution to critical ethnography in geography not only through the empirical reach and comparisons of two globalized green bean networks, but also in emphasizing the crucial role of culture in shaping production and consumption decisions and practices. In this way, she extends our knowledge of how to further integrate a Marxian analysis of the political economy of agrarian development and change with a cultural analysis of consumption and commerce, making an excellent contribution to research on the geographies of food and agriculture.[32]

In short, one does not have to like green beans to find something of value in Freidberg's book, for as Jarosz effectively conveys, *French Beans* will be of interest both to scholars working in the fields of political economy, agrarian development, and

---

29 Diener, "Review of *Seeking Asylum,*" 563.
30 Lucy Jarosz, "Review of *French Beans and Food Scares: Culture and Commerce in an Anxious Age,* by Susanne Freidberg," *Annals of the Association of American Geographers* 96, no. 1 (2006): 230–232.
31 Jarosz, "Review of *French Beans,*"231–232.
32 Jarosz, "Review of *French Beans,*" 232.

commodity chains, and with geographers and scholars using ethnographic methods in their research.

Debates in Geography as Read through Book Reviews

Book reviews are frequently read in isolation, that is, book reviews are read to determine if any given book is useful for research or teaching activities. However, book reviews often serve as platforms for scholarly debate. Indeed, the history of geography is replete with many well-known debates that play out in the form of book reviews. A well-documented encounter, for example, is that between David Harvey and Brian Berry that stemmed from the latter's review of Harvey's *Social Justice and the City*.[33] A lesser known, but equally informative debate played out in a series of papers authored by Judy Walton, Don Mitchell, and Richard Peet on the subject of landscapes, realism, and idealism in geography.[34] The exchange began innocuously enough, with book reviews of two different books, written by two different people, that appeared in two different journals. In 1993 Don Mitchell published in *The Professional Geographer* a review of Trevor Barnes and James Duncan's edited volume *Writing Worlds: Discourse, Text and Metaphor in the Representation of Landscape.* Also in that year, Richard Peet published in the *Annals of the Association of American Geographers* a review of James Duncan's scholarly monograph *The City as Text: The Politics of Landscape Interpretation in the Kandyan Kingdom.*

Peet frames his review around the debates raging between post-structuralism and Marxism that punctuated critical geography during the 1990s. He opens: "If a single tendency characterizes poststructuralist thought, it is the linguistic turn toward discourse, text, reading, and interpretation. Insights developed in linguistic

---

33 Brian J.L. Berry, "David Harvey: Social Justice and the City," *Antipode* 6, no. 2 (1974): 142–149. See also the discussion in Paul Cloke, Chris Philo, and David Sadler, *Approaching Human Geography: An Introduction to Contemporary Theoretical Debates* (New York: The Guilford Press, 1991), 37–40.
34 Richard Peet, "Review of *The City as Text: The Politics of Landscape Interpretation in the Kandyan Kingdom*, by James S. Duncan," *Annals of the Association of American Geographers* 83, no. 1 (1993): 184–187; Don Mitchell, "Review of *Writing Worlds: Discourse, Text and Metaphor in the Representation of Landscape*, edited by Trevor Barnes and James S. Duncan," *The Professional Geographer* 45, no. 4 (1993): 474–475; Judy R. Walton, "How Real(ist) Can You Get?" *The Professional Geographer* 47, no. 1 (1995): 61–65; Don Mitchell, "Sticks and Stones: The Work of Landscape (A Reply to Judy Walton's 'How Real(ist) Can You Get?')" *The Professional Geographer* 48, no. 1 (1996): 94–96; Richard Peet, "Discursive Idealism in the 'Landscape-as-Text' School," *The Professional Geographer* 48, no. 1 (1996): 96–98; and Judy Walton, "Bridging the Divide—A Reply to Mitchell and Peet," *The Professional Geographer* 48, no. 1 (1996): 98–100.

philosophy and literary theory have drawn a wide following, especially among those critical of the theoretical claims of historical materialism and the politics of socialism."[35] Peet positions Duncan as being prominent in the rejuvenation of cultural geography in the light of advances in post-structural social and literary theory. Consequently, Peet reviews *The City as Text* within these broader disciplinary trends. Here, the substance of the debate, and the content of Peet's review, is less crucial than the exchange initiated in this opening salvo. The operative element is Peet's charge that Duncan appears to bring into geography "a new form of idealism."[36] Peet concludes his review with the following summation:

> This is a well-written, fastidiously researched book by an author who has steeped himself in Sri Lankan history and culture. In terms of method, however, my conclusion, based on a critical reading of the book, has to be that the intertextual version of poststructural cultural theory is but a new form of that older approach to history and geography: elitist idealism.[37]

In sum, Peet provides a critical review of Duncan's scholarship, with a particular focus on the theories and methods employed by Duncan. And based on this reading, Peet makes a bold claim regarding the status of critical/radical geography.

Mitchell's review of Barnes and Duncan's edited volume, *Writing Worlds*, also places the book within the disciplinary debates of cultural geography, post-structuralism, and Marxism. Mitchell begins: "This is a frustrating book saved only by the contributors who do not live up to the task given to them by the editors. The collection is frustrating because the editors establish a theoretical position —which they claim binds the collection as a whole—that is untenable at best and politically stultifying at worst."[38] As an opening, Mitchell deftly addresses the strengths (and weaknesses) of edited volumes. Well-structured edited volumes come together as coherent wholes, with a high degree of continuity within and between chapters. Here, however, Mitchell directs attention to the conceptual glue that should otherwise bind together a collection of disparate chapters. For Mitchell, the problem is in part structural, that is, "in their introductory essay Barnes and Duncan establish a curious theoretical universe."[39] In Mitchell's review, however, he finds that the contributions to the book are lacking.

More precisely, Mitchell determines that the individual chapters fail to adhere to the conceptual foundation established by Barnes and Duncan in their introduc-

---

35 Peet, "Review of *The City as Text*," 184.

36 Peet, "Review of *The City as Text*," 186.

37 Peet, "Review of *The City as Text*," 187.

38 Mitchell, "Review of *Writing Worlds*," 474.

39 Mitchell, "Review of *Writing Worlds*," 474.

tion. In making this claim, Mitchell does not just provide a summation of individual chapters but also evaluates their merits according to the collection as a whole. Mitchell begins, "Luckily, most of the contributions claim the more important ground of trying to understand how 'reality' and language interact. Seven of the essays explicitly seek to understand how particular discourses abstract from the 'real' and are in turn made even more 'real' by their material construction."[40] Throughout his review, Mitchell strikes a balance between the singular contributions and how they relate to the overall project. For example, Mitchell writes,

> McGreevy shows how discourses of death around Niagara Falls are structured by other discourses *and* the physicality of the falls itself—though the relations between these two realms is highly complex. While McGreevy tries to shoehorn his material into the bounds established by Barnes and Duncan, who write that 'travelers' texts are about other texts and not some pristine falls' (p. 4), the shear natural power of the falls keeps imposing itself on McGreevy's narrative.[41]

Here, Mitchell draws together material presented in Barnes and Duncan's introductory chapter and how this rests uneasily with McGreevy's contribution. In short, Mitchell never loses sight of (1) the book as a purported coherent whole; and (2) how the edited volume fits within the wider debates of cultural geography and landscape studies.

Without doubt, Peet and Mitchell both make bold claims vis-à-vis geography through their book reviews. This, we have seen, is not uncommon and is something many editors, including those affiliated with the *AAG Review of Books*, now explicitly ask of reviewers. What is remarkable for our immediate purpose is that the debate was continued by Judy Walton by way of a commentary published in *The Professional Geographer.* Walton opens: "It's funny how old (and tiresome?) debates in geography never die, they just find new battlefields. Ever since structural Marxism within geography was first accused by Duncan and Ley (1982) of harboring idealist roots, a line of attack has formed that throws back this philosophical pejorative at the accusers."[42] With this salvo, Walton traces the debate back at least to the early 1980s and asserts that the present dispute is merely old wine in new bottles. Walton continues, "Two book reviewers, Don Mitchell (1993) in *The Professional Geographer* and Richard Peet (1993) in the *Annals of the Association of American Geographers*, have recently leveled charges of 'idealism' against, respectively, Barnes and Duncan's (1992) *Writing Worlds* and Duncan's (1990) *The City as Text*

---

40 Mitchell, "Review of *Writing Worlds*," 474.
41 Mitchell, "Review of *Writing Worlds*," 474.
42 Walton, "How Real(ist)," 61.

for their claim that culture and/or landscape can be seen as text(s)."[43] In the remainder of her commentary, Walton *reviews* the *reviews* of both Mitchell and Peet!

My sustained engagement with the Mitchell, Peet, and Walton exchange is not to wander into the theoretical thickets of realism and idealism in the geographic study of landscapes. Instead, I use this written exchange to illustrate how book reviews can and do serve as effective outlets for geographic debate. That these debates continue in other forms should not be lost. Indeed, it is worth quoting Mitchell's response to Walton:

> It is a rare honor when one's *book review* (Mitchell 1993b) is taken seriously enough to merit a full critique in a major journal. I am pleased that the point I was trying to make, that we need to much more carefully think through the relations of discourse and reality embodied in landscapes, came through clearly in such a short review and is worthy of careful attention.[44]

Mitchell's statement, concisely, holds for writing book reviews in general: the function of a well-crafted review is to inform and stimulate; it is to engage purposefully with the subject-matter and to not lose sight of the forest (geography) from the trees (individual books).

Some Final Thoughts on Writing Book Reviews

Book reviews do not always receive the respect they are due. As Jarvis bemoans, "In a hierarchy of publication genre spanning online and hard copy books, essays and journal articles, it is tempting to dismiss the humble book review as something of little consequence."[45] And yet, book reviews serve a vital function in academia. We routinely turn to book reviews to help decide book(s) to adopt for our courses. We also read reviews to see if any given book might help in our own research. And we read book reviews to gain a better understanding and appreciation of trends and trajectories of geography and other fields of study.

---

43 Walton, "How Real(ist)," 61.
44 Mitchell, "Sticks and Stones," 94.
45 Jarvis, "Book Reviews," 16.

Chapter 7 Encyclopedia Entries

Encyclopedia entries differ from journal articles and book chapters in several distinct ways. Most obviously, entries are shorter. Depending on the journal, published articles in geography typically range between 5,000 and 12,000 words, with the majority spanning between 6,000 and 8,000 words. Encyclopedia entries may surpass 3,000 words but often run between 1,500 and 2,000 words. In fact, some entries may be as brief as 500 words. If you're used to writing 5,000-word journal articles, you will quickly realize how little space there is in a 500-word entry.

Unlike (most) research articles and book chapters, encyclopedia entries do not present original research. Instead, entries provide a concise overview of subject, usually supported by secondary sources. Heuristically, I find it helpful to compare abstracts with encyclopedia entries. If the former involves the distillation of a particular research project, the latter constitutes the condensation of a specific research topic. In other words, encyclopedia entries provide the essence of a topic.

Encyclopedia entries should be comprehensive, that is, able to synthesize the complexities of a topic. However, entries are not exhaustive. Unlike book-length treatments of a topic, which can extend for over 80,000 words, encyclopedia entries are brief and necessarily succinct. Indeed, this brevity often constitutes the major hurdle when writing encyclopedia entries. A general rule of thumb: when writing these academic pieces, clarity is crucial. There is little scope for flowery language or other literary devices. Almost without exception, allegory and imagery are to be avoided, for the point of the entry is to communicate plainly the fundamental substance of the subject-matter.

It's helpful, at this point, to position encyclopedia entries within a wider academic context. Many of us are well familiar with encyclopedias. When I was young, I would take from library bookshelves heavy, embossed encyclopedias to help when writing essays or term papers. Now, encyclopedias often are in digital form—but their purpose largely remains the same. Encyclopedias provide the nuts-and-bolts of a topic in non-specialized language. However, there are also many specialized encyclopedias that comprise entries on narrowly circumscribed topics. Before writing an entry, it is imperative to fully acknowledge and to be aware of the intended audience. Simply put, certain key words, concepts, and theories will or will not require elaboration, depending on the background of the intended audience.

The clarification of key words, concepts, and theories relates to another key distinction between encyclopedia entries and journal articles (especially) and (some) book chapters. When writing encyclopedia entries, it is possible—and desir-

https://doi.org/10.1515/9783111189727-009

able!—to cross-reference other entries that define the key words, concepts, and theories under question. For example, in writing an encyclopedia entry on the concept of the 'Responsibility to Protect' (R2P), I might discuss the limitations to genocide prevention efforts at the United Nations. As an encyclopedia entry on R2P, I don't have to define 'genocide' or the 'United Nations,' provided these two terms are defined elsewhere.

Encyclopedia entries are (usually) required to adhere to a particular framework or template. Journal articles and book chapters, we have seen, often are expected to conform to specific stylistic conventions. However, with encyclopedia entries, conformity among entries is often the coin of the realm. This is, as indicated, because encyclopedias are meant to convey clearly and concisely the essence of a subject for an intended audience. In fact, it is this conformity that makes possible the *intertextual* strengths of an encyclopedia, for example to cross-reference entries on the 'Responsibility to Protect' with entries on 'genocide.'

The Format of Encyclopedia Entries

It is neither possible nor advisable to provide a universal template or format for writing encyclopedia entries, for the simple reason that when invited to write an entry, a style guide will be provided. That said, let's take a look at the guidelines provided for *The International Encyclopedia of Geography: People, the Earth, Environment, and Technology*, edited by Douglas Richardson and colleagues. An on-line encyclopedia, the IEG provides short entries on key concepts and extended explorations of major topics. Entries are intended to appeal to a range of geographers, from advanced undergraduates to seasoned scholars. Guidelines suggest that entries include the topic's intellectual and social context; major dimensions of the topic; changes over time in the topic and its treatment; current emphases (and debates) in work on the topic in research and theory; and future directions in research, theory, and methodology. All entries, lastly, are to include abstracts, key words, cross-references to other entries in the encyclopedia, references cited in the main entry, and suggestions for further reading. By way of illustration, I was asked to contribute an entry to this encyclopedia on the topic of genocide.[1]

My entry is approximately 3,500 words and consists of three sections: (1) introduction; (2) genocide as a geographic process; and (3) genocide and geospatial tech-

---

[1] James A. Tyner, "Genocide," in *The International Encyclopedia of Geography*, edited by Douglas Richardson, Noel Castree, Michael F. Goodchild, Audrey Kobayashi, Weidong Liu, and Richard Marston (New York: John Wiley & Sons, 2021): https://doi.org/10.1002/9781118786352.wbieg2098.

nologies. In the introduction I briefly describe the origins of the term and discuss the various challenges and extensions of the term. I highlight that the topic of genocide has been studied most prominently by legal scholars, historians, and anthropologists; but that geographers have made substantial contributions. In the second section I provide a discussion of how genocide has been conceived as a geographic process. Specific reference is made to the scholarly contributions of geographers who elaborate on this notion. Lastly, I provide an overview of how geospatial technologies, including Geographic Information Systems and remote sensing, are employed both by geographers and non-geographers in the study of genocide. Specific examples are used for illustrative purposes.

By way of comparison, consider the entry on "Home," co-authored by Jon May and Paul Cloke, published in the *Encyclopedia of Human Geography*.[2] Their entry is made up of five paragraphs; none contains sub-headings. May and Cloke begin with a short, declarative sentence: "The home is both a material place—a building, often with a garden or yard attached, located in a particular neighborhood—and a space in which identities and meanings are constructed."[3] As a geographer, I find this opening particularly well conceived, in that it immediately introduces several concepts that are central to our discipline, notably place, space, location, and identities. In addition, it underscores both the material reality of home and the discursive (ideological) meanings associated with the term home. May and Cloke continue: "Over the years, geographers have assumed the home to be a site of unchanging and stable social geographies, but more recently this assumption has been challenged on a number of fronts as conventional meanings of home have been scrutinized and deconstructed. As a result, the home has become a more fluid and contested space."[4]

There's considerable information packed into these two sentences. First, the authors indicate that (1) geographers have long studied 'homes' within the context of social geography, but (2) traditional ways of conceiving 'home' have been questioned. The last sentence foreshadows that the meaning—and, by implication, the lived experience—of home has changed. As a reader, I would expect, therefore, that the subsequent paragraphs of the entry clarify these transformations.

May and Cloke do not let us down. The second paragraph details how the home, traditionally, has been conceived in terms such as shelter, hearth, heart, privacy, roots, abode, and paradise. Furthermore, May and Cloke write that "many of

2 Jon May and Paul Cloke, "Home," in *Encyclopedia of Human Geography*, edited by Barney Warf (Thousand Oaks, CA: SAGE Publications, 2006), 225–226.
3 May and Cloke, "Home," 225.
4 May and Cloke, "Home," 225.

these meanings have also been related to the scale of nationhood, indicating the inclusions (and exclusions) of a home country."[5] In doing so, May and Cloke connect the entry on home to several other key concepts in geography, notably scale, nation, and (implicitly) sovereignty. The subsequent three paragraphs detail three specific challenges to the traditional assumptions of home and clarify how geographers have forwarded alternative ways of understanding and analyzing home as a concept. They explain, first, that "geographers have recognized home as a place of labor" and that "feminist geographers in particular have critiqued any formula that separates out the 'private' space of home from the 'public' space of work." Second, they highlight that "home has been recognized as a space of oppression" and that "Geographers have provided evidence of how expressions of male domination, dysfunctional 'family' circumstances (e. g., the introduction of new and unsympathetic stepparents), or issues relating to alternative sexuality can lead directly to oppression of young people in their home, and teenagers leaving home are accounting for increasing proportions of the homeless population." Third, they explain that "home is a space of negotiation and contestation" and that "Geographers have emphasized that the home is a site where identity and meaning are constructed through consumption and where connections are made with global and local discourses about the home."[6]

Let's step back a moment and really think about what May and Cloke accomplish in this entry. Following a crisp, cogent introduction to the subject-matter, they organize their manuscript in a way that is both coherent and logical. Home, traditionally, has been understood in one way; but in recent years, geographers have challenged prior assumptions. Three specific challenges are identified—each comprising a unique paragraph—and specific examples of how geographers have forwarded these challenges are included. Throughout the entry, several concepts developed and expanded upon by geographers are introduced. Apart from the aforementioned terms of place, space, location, scale, nation, and sovereignty, we have public and private spheres, labor, violence, fear, family, homelessness, emotions, consumption, globalization, resistance, time–space boundaries, and performance. What more can one ask from a short entry written to introduce readers to a particular topic?

Alison Mountz, writing also in the *Encyclopedia of Human Geography*, provides another helpful example of writing an effective encyclopedia entry, in this case on the complex topic of the 'state.'[7] Following a brief introduction, Mountz' contribu-

5 May and Cloke, "Home," 225.
6 May and Cloke, "Home," 225–226.
7 Alison Mountz, "State," in *Encyclopedia of Human Geography*, edited by Barney Warf (Thousand Oaks, CA: SAGE Publications, 2006), 460–462.

tion is made up of four sections: (1) history of the state; (2) contemporary typologies; (3) epistemologies; and (4) geography and the state. She begins, as do May and Cloke, with a straightforward definition: "The state refers to a geographic area delineated by national borders, the population that inhabits this territory, and the political unit and institutions that govern the social and economic relationships among people."[8] And, similar to May and Cloke, several geographic terms and concepts are introduced. She clarifies next that the term 'state' can describe three levels of government, namely federal, state/provincial, and local. In addition, at each level, the term 'state' refers also to the "assemblage of institutions, networks, administrative functions, and people organized by bureaucratic organizations."[9] In this way, she highlights the differences in terminology of state, nation-state, and country, and explains that confusion over terminology abounds.

Beyond the definitional criteria (and confusion) of the state, Mountz concludes the introduction by noting that "the state has a lively conceptual existence in the social sciences" and that there "likely exist more theories of the state than kinds of state."[10] This is a wonderful example of foreshadowing, in that Mountz both introduces the existence of *typologies* of states and of competing *epistemologies* of the state. Also, Mountz identifies that "Geography and geographers have engaged unevenly with theories of the state ever since the inception of the discipline" and that while political geographers have long participated in these debates, there has been "a recent surge of interest across sub-disciplinary fields."[11] We can infer from these brief remarks that the geographic study of states is a vibrant field, and Mountz captures much of this excitement. In doing so, Mountz draws the reader into the topic.

Mountz' section on the history of the state provides much-needed background on the origins of the state as concept, its relation to agrarian societies, feudalism, empires, and capitalism. She highlights how the modern 'nation-state' differs from other forms of political organization and the increasing homogenization of states as political entities during the twentieth century. This section is punctuated with many key concepts: territory, power, militarism, supra-state institutions, decolonization, social movements, ethnic separatism, and globalization. Lastly, Mountz follows with a brief sub-section on contemporary typologies—an element foreshadowed in the introduction. Here, Mountz explains that "All states exercise power

8 Mountz, "State," 460.
9 Mountz, "State," 460.
10 Mountz, "State," 460.
11 Mountz, "State," 460.

through sovereignty, but they do so with various methods of governance."[12] To that end, she introduces two main ways in which states are classified.

Beyond simple classifications, Mountz clarifies that geographers have relied on diverse epistemological approaches to the state, including Marxism, feminism, and post-structuralism. And while diverse in their approach, Mountz adds, all epistemologies hold "in common the project of spatializing theories of the state."[13] Effectively, Mountz represents the theoretical study of the state as a microcosm of geography. Indeed, Mountz wonderfully captures what I believe lies at the core of our discipline, that is, we can work from many different perspectives, such as logical positivism, Marxism, feminism, or post-structuralism, but there remains *something* that we hold in common, something that enables geographers regardless of epistemological stripe to come together. This does not mean that geographers are always in agreement. However, it is possible for geographers to constructively engage in these differences and, in doing so, call attention to elements not previously addressed.

In her closing section, Mountz highlights more explicitly the interconnections of geography and the state. Here, Mountz reaffirms that "the state always has been among the topics central to the work of political geographers" but that in recent decades alternative approaches to the state have appeared. Mountz identifies the source of these extensions to the study of the state, for example the influence of globalization and neoliberalism on the system of nation-states. She identifies, also, the influence of key theorists, such as Henri Lefebvre and Michel Foucault, and how this has redirected the geographic study of the state away from "states as containers" and toward the production of states at various scales. Lastly, Mountz details the flourishing of insights derived from post-colonial and feminist perspectives, and how these latter influences have "brought histories of violence to bear on understandings of the state."[14]

In short, Mountz provides a clearly written, well-structured entry that deftly communicates the many ways the 'state' has been defined, operationalized, and studied in geography. She provides a brief but detailed history of the term, followed by competing typologies and epistemologies that frame the study of the state. She acknowledges how our understanding of the state has transformed, influenced both by external 'sources' (such as the rise of neoliberal governance) and internal 'sources' (e.g., the incorporation of post-colonial epistemologies into geography).

12 Mountz, "State," 461.
13 Mountz, "State," 461.
14 Mountz, "State," 462.

Some Final Thoughts on Writing Encyclopedia Entries

In recent years, the number of encyclopedias, both within geography and in other disciplines, has increased greatly. Indeed, with on-line publishing, there is a veritable cottage industry of encyclopedias. This is not necessarily a bad thing and I am not passing judgment. In fact, rather than weigh the advantages and disadvantages of encyclopedias in geography, I want to call attention to the practice of writing encyclopedia entries for alternative purposes. In short, I want to consider more broadly why anyone would want to adopt this writing *format*.

From the outset, I've emphasized that writing is a skill; and that two banal but crucial training exercises are reading and writing. To that end, and especially for researchers new to a particular topic, I encourage writing encyclopedia entries for one's own personal scholarly development. Traditionally, we acquaint ourselves with research projects through the dreaded 'literature review' assignment. Here, students and even more experienced writers often fall into the trap of providing a litany of disparate scholarship. At the risk of hyperbole, the typical literature review read something like "In 2009 Smith argues that ...; in 2012 Rodriquez argues that ...; and in 2018 Li argues that ...." Such a presentation is organized chronologically; but disorganized in thought. What are the main trends? What are the competing perspectives? What are the specific contributions geographers have made, and why? The encyclopedia entry, as a *form of writing*, provides an ideal technique to gather one's thoughts and present a cogent and coherent overview of a topic. To that end, writers should think about writing encyclopedia entries—not necessarily for publication—but as an exercise to (1) acquire a deeper knowledge and understanding of a topic; and (2) cultivate their ability to effectively synthesize and communicate the essence of a subject in written form.

Chapter 8 Commentaries

Briefly stated, a commentary is a short manuscript that communicates an opinion or explanation on a particular subject. In geography, scholars often write commentaries on current events, for example an armed conflict or policy change. Geographers also write commentaries to provide clarity on key concepts or theories, such as the 'Anthropocene' or 'post-humanism,' or in response to on-going debates. Commentaries may include primary data or primarily express one's personal view of a particular subject.

Many academic outlets in geography, including several of the leading scholarly journals, provide forums for scholars to write commentaries. In addition, with recent advances in digital publishing, it is possible to write commentaries as blogs—perhaps in connection with newsletters, journals, presses, and institutional organizations. *The Geographical Journal*, for example, allows contributors to submit commentaries, defined here as shorter pieces (2,000 to 4,000 words) reflecting on an event or topic of contemporary relevance to policy, public, and/or academic debates.[1] Conversely, *The Professional Geographer* publishes commentaries only on articles published in the journal. As stated in the journal guidelines, "commentaries should focus on a specific article and be written in a style and tone that is professional, scholarly and concise, in less than 2,000 words including references. Commentaries should be submitted within one year after publication of the original article." And to underscore the limited scope for commentaries in *The Professional Geographer*, guidelines state clearly that "commentaries on themes or special issues will not be published."[2]

A variant of conventional commentaries is found in *Human Geography*. As a self-styled radical journal, *Human Geography* includes a forum for shorter, original contributions. Billed as 'contentions' as opposed to 'commentaries,' authors are encouraged to

> give a perspective on current or past events or intervene in topical debates. This section of the journal seeks to bridge the divide between academic and more journalistic writing. It seeks to apply critical thinking in human geography to real world situations with contemporary rele-

---

1 *The Geographical Journal*, "Author Guidelines," available at https://rgs-ibg.onlinelibrary.wiley.com/hub/journal/14754959/homepage/forauthors.html#TYPES (accessed January 23, 2023).

2 *The Professional Geographer*, "The Professional Geographers Guidelines for Commentaries," available at https://www.aag.org/professional_geographer_guidelines_for_commentaries-3/ (accessed January 23, 2023).

https://doi.org/10.1515/9783111189727-010

vance, for example those related to environmental issues, race, gender, sexuality, war and conflict, workers' struggles, or other movements against injustice and oppression.[3]

Writing Commentaries

A commentary written for an academic outlet, including journals, may be invited by an editor or self-submitted by an author or authors. In the latter case, the author(s) will contact the journal editor and query if the prospective subject is of interest. Regardless, the commentary (typically) will have to adhere to a specific format (see above) and will undergo a review process. Unlike refereed articles, the review process may be abbreviated, that is, undertaken by the editor and/or members of the editorial review. This form of review is necessary, essentially, because the commentary often is centered around personal opinions, written to be provocative. It would defeat the purpose of a commentary if a reviewer could reject the piece over scholarly disagreement. That said, some review is necessary, to ensure (hopefully) that certain facts are not mis-represented.

Apart from pre-set guidelines, commentaries ran the gamut in terms of style, substance, and tenor. Some commentaries are straightforward, not unlike encyclopedia entries. In this format, commentaries provide clear (objective) discussions of specific events, issues, policies, or concepts. Unlike encyclopedia entries, however, in these commentaries the author(s) may include their own opinions or reflections on the subject. In a commentary published in *Environment and Planning D: Society and Space*, for example, Sara Fregonese, Diana Martin, and Adam Ramadan remarked on a June 2009 speech delivered by then-United States President Barack Obama.[4] Noting that Obama's speech seemingly articulated a new vision of US geopolitics, especially as this related to the Muslim world, Fregonese, Martin, and Ramadan asked a series of provocative questions that emanated from Obama's speech. In this commentary, the authors were able to use a specific speech as a springboard to reflect on the wider implications of US geopolitics—topics of significant interest to geographers. Likewise, Louise Amoore and Marieke de Goede called attention to Western responses to the 2010 floods in Pakistan to consider more deeply other forms of violence, including the use of drone attacks by the

---

3 *Human Geography*, "Submission Guidelines," available at https://journals.sagepub.com/author-instructions/HUG#Articletypes (accessed January 23, 2023).
4 Sara Fregonese, Diana Martin, and Adam Ramadan, "Commentary: The New Geopolitics of Responsibility in Barack Obama's Cairo Speech," *Environment and Planning D: Society and Space* 27 (2009): 951–955.

United States in the so-called War on Terror. Commenting on how seemingly non-violent targetings such as financial black-listings impinge on humanitarian work conducted in the region, Amoore and de Goede assert that charities, especially Islamic faith-based organizations, face an impossible terrain of operations and have genuine difficulty doing their work.[5]

For comparison, let's look at a recent commentary I published with Stian Rice in *Human Geography*. In this instance, our piece appeared in that journal's "Contentions" section and we chose to intentionally write our piece as a polemic. The subject at hand was a recent announcement that engineers had developed a technique to 'reanimate' a dead spider and to subsequently manipulate the *necrobot*—their term—to perform labor. We determined that *Human Geography*'s radical edge was an excellent venue to confront the vitality of capitalism. Indeed, guidelines mention specifically that "although contributions should be intellectually rigorous and may be informed by the author's previous research, contentions are generally more focused on topical issues and can be more polemical in tone than a research article."[6]

Freed from the constraints of any semblance of objectivity, we introduce our commentary with the Witches' scene in Act IV of Shakespeare's "Macbeth," ending with the lines "By the pricking of my thumbs/something wicked this way comes." We then announce our main contention, writing that "Into the bubbling cauldron of capitalism, scientists have recently added a spider. It turns out that spiders make excellent little grippers, ideal for picking up and placing tiny parts in a wide range of manufacturing applications. Of course, not all spiders will do. Only dead ones."[7] Subsequently, we argue that the killing and reanimating of spiders raises several issues of concern to geographers, chiefly the fetishization of violence, the subsumption of life and death to circuits of capital, and the contradictions inherent in necrocapitalism. In terms of composition, Stian and I draw heavily on a number of literary techniques, including the liberal use of metaphors and intertextuality. We are able to do so, in large part, because of the format offered by *Human Geography*.

---

5 Louise Amoore and Marieke de Goede, "Commentary: Risky Geographies: Aid and Enmity in Pakistan," *Environment and Planning D: Society and Space* 29 (2011): 193–202; at 200.

6 *Human Geography*, "Submission Guidelines," available at https://journals.sagepub.com/author-instructions/HUG#Articletypes (accessed January 23, 2023).

7 James A. Tyner and Stian Rice, "Along Came a Spider ... and Capitalism Killed It," *Human Geography* (2022), https://doi.org/10.1177/19427786221140843 (accessed January 23, 2023).

Some Final Thoughts on Writing Commentaries

Novice writers are probably not in a position to write commentaries. Editors typically will solicit commentaries from authors who have more in-depth knowledge of the particular topic and are in a position to offer a new or unique viewpoint or to reflect critically on a topic. That said, we all have opinions on any number of subjects and are thus qualified to 'speak our mind.' As someone just getting started, there's no reason why you can't think about writing commentaries. Indeed, in some situations editors might expressly seek out graduate students and early-career academics for their insight and personal views. Keep in mind, though, that writing commentaries in an academic setting is not the same as posting your thoughts on a blog. Commentaries in geography journals are (often) refereed publications, subject to peer review. This doesn't mean you can't be creative or polemical; in fact, some of the most provocative commentaries are imaginatively written. It does mean, however, that certain standards are followed, including those related to professionalism, truthfulness, and honesty. To be provocative is not the same as being deceitful or harmful.

Works Cited

AAG Review of Books, "Guidelines for Reviewers," available at http://www.aag.org/AAGRB_Reviewer_ Guidelines (accessed April 14, 2021).

Aitken, Stuart C. and Leo E. Zonn (eds), *Place, Power, Situation, and Spectacle: A Geography of Film* (Lanham, MD: Rowman & Littlefield, 1994).

American Geographical Union, "Checklist for Submitting a Paper to an AGU Journal," available at https://www.agu.org/Publish-with-AGU/Publish/Author-Resources/New-Manuscript-Checklist (accessed April 13, 2021).

Amoore, Louise and Marieke de Goede, "Commentary: Risky Geographies: Aid and Enmity in Pakistan," *Environment and Planning D: Society and Space* 29 (2011): 193–202.

Annals of the American Association of Geographers, available at https://www.tandfonline.com/journals/ raag21 (accessed April 16, 2021).

Askins, Kye, "Being Together: Everyday Geographies and the Quiet Politics of Belonging," *ACME: International E-Journal for Critical Geographies* 14, no. 2 (2015): 461–469.

Askins, Kye and Matej Blazek, "Feeling Our Way: Academia, Emotions and a Politics of Care," *Social & Cultural Geography* 18, no. 8 (2017): 1086–1105.

Bailey, Cathy, Catherine White, and Rachel Pain, "Evaluating Qualitative Research: Dealing with the Tension between 'Science' and 'Creativity,'" *Area* 31, no. 2 (1999): 169–183.

Barnes, Jessica, *Cultivating the Nile: The Everyday Politics of Water in Egypt* (Durham, NC: Duke University Press, 2014).

Barnes, Trevor, "The Future of Research Monographs: An International Set of Perspectives," in Kevin Ward, Ron Johnston, Keith Richards, Matthew Gandy, Zbigniew Taylor, Anssi Paasi, Roddy Fox, Margarita Serje, Henry Wai-chung Yeung, Trevor Barnes, Alison Blunt, and Linda McDowell, "The Future of Research Monographs," *Progress in Human Geography* 33, no. 1 (2009): 101–126.

Baxter, Jamie and John Eyles, "Evaluating Qualitative Research in Social Geography: Establishing 'Rigour' in Interview Analysis," *Transactions of the Institute of British Geographers* 22, no. 4 (1997): 505–525.

Berg, Lawrence D., Edward H. Huijbens, and Henrik Gutzon Larsen, "Producing Anxiety in the Neoliberal University," *The Canadian Geographer/Le Géographe canadien* 60, no. 2 (2016): 168–180.

Berry, Brian J.L., "David Harvey: Social Justice and the City," *Antipode* 6, no. 2 (1974): 142–149.

Bickham, Jack M., *Scene and Structure* (Cincinnati, OH: Writer's Digest Books, 1993).

Bledsoe, Adam and Willie Jamaal Wright, "The Anti-Blackness of Global Capital," *Environment and Planning D: Society and Space* 37, no. 1 (2019): 8–26.

Blocken, Bert, "10 Tips for Writing a Truly Terrible Journal Article," *Elsevier Publishing Campus*, March 1, 2017, available at http://www.urbanphysics.net/Elsevier_Publishing_Campus_Blocken_2017_PDF. pdf (accessed January 19, 2023).

Blunt, Alison, "Books and Individual Publication Strategies," in Kevin Ward, Ron Johnston, Keith Richards, Matthew Gandy, Zbigniew Taylor, Anssi Paasi, Roddy Fox, Margarita Serje, Henry Wai-chung Yeung, Trevor Barnes, Alison Blunt, and Linda McDowell, "The Future of Research Monographs," *Progress in Human Geography* 33, no. 1 (2009): 120–121.

Blunt, Alison and Catherine Souch (eds), *Publishing in Geography: A Guide for New Researchers* (London: Royal Geographical Society, 2008).

Bondi, Liz and Mona Domosh, "On the Contours of Public Space: A Tale of Three Women," *Antipode* 30, no. 3 (1998): 270–289.

https://doi.org/10.1515/9783111189727-011

Boschmann, E. Eric and Emily Cubbon, "Sketch Maps and Qualitative GIS: Using Cartographies of Individual Spatial Narratives in Geographic Research," *The Professional Geographer* 66, no. 2 (2014): 236–248.

Bourne, L.S., "On Writing and Publishing in Human Geography: Some Personal Reflections," in *On Becoming a Professional Geographer*, edited by Martin S. Kenzer (Caldwell, NJ: The Blackburn Press, 2000), 100–112.

Bracken, Louise J. and Alastair Bonnett, "Research Articles," in *Publishing in Geography: A Guide for New Researchers*, edited by Alison Blunt and Catherine Souch (London: Royal Geographical Society, 2008), 4–12.

Bradshaw, Matt and Elaine Stratford, "Qualitative Research Design and Rigour," in *Qualitative Research Methods in Human Geography*, edited by Iain Hay (Oxford: Oxford University Press, 2005), 67–76.

Bradshaw, Michael J. and Rochelle Lieber, "Review Essays," in *Publishing in Geography: A Guide for New Researchers*, edited by Alison Blunt and Catherine Souch (London: Royal Geographical Society, 2008), 14–16.

Broadway, Michael J., Robert Legg, and Teresa Bertossi, "North American Independent Coffeehouse Culture: A Comparison of Seattle with Vancouver," *GeoJournal* 85 (2020): 1645–1662.

Bromley, Ray, "Review of *Placing Animals: An Introduction to the Geography of Human–Animal Relations*, by Julie Urbanik," *The AAG Review of Books* 2, no. 4 (2014): 133–135.

Brown, J. Christopher, Lisa Rausch, and Verônica Gronau Luz, "Toward a Spatial Understanding of Staple Food and Nonstaple Food Production in Brazil," *The Professional Geographer* 66, no. 2 (2014): 249–259.

Brunn, Stanley D., "Personal and General Publishing Policies of Geographers," *Terra* 99, no. 3 (1987): 155–165.

Brunn, Stanley D., "The Manuscript Review Process and Advice to Prospective Authors," *The Professional Geographer* 40, no. 1 (1987): 8–14.

Brunn, Stanley D., Maureen Hays-Mitchell, Donald J. Zeigler, and Jessica K. Graybill, *Cities of the World: Regional Patterns and Urban Environments*, 6th edition (Lanham, MD: Rowman & Littlefield, 2016).

Bruno, Tianna and Cristina Faiver-Serna, "More Reflections on a White Discipline," *The Professional Geographer* 74, no. 1 (2022): 156–161.

Burlingame, Katherine, "Where are the Storytellers? A Quest to (Re)enchant Geography through Writing as Method," *Journal of Geography in Higher Education* 43, no. 1 (2019): 56–70.

Butler, David R., "Conducting Research and Writing an Article in Physical Geography," in *On Becoming a Professional Geographer*, edited by Martin S. Kenzer (Caldwell, NJ: The Blackburn Press, 2000), 88–99.

Cameron, Jenny, Karen Nairn, and Jane Higgins, "Demystifying Academic Writing: Reflections on Emotions, Know-How and Academic Identity," *Journal of Geography in Higher Education* 33, no. 2 (2009): 269–284.

Card, Orson Scott, *Characters and Viewpoint* (Cincinnati, OH: Writer's Digest Books, 1988).

Card, Orson Scott, *How to Write Science Fiction and Fantasy* (Cincinnati, OH: Writer's Digest Books, 1990).

Caretta, Martina Angela, Danielle Drozdzewski, Johanna Carolina Jokinen, and Emily Falconer, "'Who Can Play *This* Game?' The Lived Experiences of Doctoral Candidates and Early Career Women in the Neoliberal University," *Journal of Geography in Higher Education* 42, no. 2 (2018): 261–275.

Castree, Noel, *Nature* (New York: Routledge, 2005).

Castree, Noel, "Research Assessment and the Production of Geographical Knowledge," *Progress in Human Geography* 30, no. 6 (2006): 747–782.

Chiarella, Tom, *Writing Dialogue* (Cincinnati, OH: Story Press, 1998).

Cloke, Paul, Chris Philo, and David Sadler, *Approaching Human Geography: An Introduction to Contemporary Theoretical Debates* (New York: The Guilford Press, 1991).

Coakley, Corrine, Mandy Munro-Stasiuk, James A. Tyner, Sokvisal Kimsroy, Chhunly Chhay, and Stian Rice, "Extracting Khmer Rouge Irrigation Networks from Pre-Landsat 4 Satellite Imagery Using Vegetation Indices," *Remote Sensing* 11 (2019): 2397.

Colucci, Alex R., "Geographies of Capital Punishment: New Directions and Interventions," *Geography Compass* 14 (2020): e12548.

Colucci, Alex R., James A. Tyner, Mandy Munro-Stasiuk, Stian Rice, Sokvisal Kimsroy, Chhunly Chhay, and Corrine Coakley, "Critical Physical Geography and the Study of Genocide: Lessons from Cambodia," *Transactions of the Institute of British Geographers* 46, no. 2 (2021): 780–793.

Creswell, John W., *Qualitative Inquiry and Research Design: Choosing Among Five Traditions* (Thousand Oaks, CA: SAGE Publications, 1998).

Cross, Cate and Charles Oppenheim, "A Genre Analysis of Scientific Abstracts," *Journal of Documentation* 62, no. 4 (2006): 428–446.

Davidson-Arnott, Robin G.D. and Bernie O. Bauer, "Aeolian Sediment Transport on a Beach: Thresholds, Intermittency, and High Frequency Variability," *Geomorphology* 105 (2009): 117–126.

Davis, Mark A. and Bernd Blossey, "Edited Books: The Good, the Bad, and the Ugly," *Bulletin of the Ecological Society of America* 92, no. 3 (2011): 247–250.

Davis, Sasha, *The Empires' Edge: Militarization, Resistance, and Transcending Hegemony in the Pacific* (Athens, GA: University of Georgia Press, 2015).

De Leeuw, Sarah and Harriet Hawkins, "Critical Geographies and Geography's Creative Re/turn: Poetics and Practices for New Disciplinary Spaces," *Gender, Place & Culture* 24, no. 3 (2017): 303–324.

Dear, Michael, "Review of *Set the Night on Fire: L.A. in the Sixties*, by Mike Davis and Jon Wiener," *The AAG Review of Books* 8, no. 4 (2020): 196–198.

DeLyser, Dydia, "Teaching Graduate Students to Write: A Seminar for Thesis and Dissertation Writers," *Journal of Geography in Higher Education* 27, no. 2 (2003): 169–181.

DeLyser, Dydia, "Writing Qualitative Geography," in *The SAGE handbook of qualitative geography*, edited by Dydia DeLyser, Steve Herbert, Stuart Aitken, Mike Crang, and Linda McDowell (Thousand Oaks, CA: SAGE, 2010), 341–358.

DeLyser, Dydia, "Writing's Intimate Spatialities: Drawing Ourselves *to* our Writing in Self-Caring Practices of Love," *EPA: Economy and Space* 54, no. 2 (2022): 405–412.

DeLyser, Dydia and Harriet Hawkins, "Introduction: Writing Creatively—Process, Practice, and Product," *Cultural Geographies* 21, no. 1 (2014): 131–134.

DeLyser, Dydia, Steve Herbert, Stuart Aitken, Mike Crang, and Linda McDowell (eds), *The SAGE Handbook of Qualitative Geography* (Thousand Oaks, CA: SAGE Publications, 2009).

Dempsey, Jessica, *Enterprising Nature: Economic, Markets, and Finance in Global Biodiversity Politics* (New York: John Wiley & Sons, 2016).

Denzin, Norman K. and Yvonna S. Lincoln (eds), *Collective and Interpreting Qualitative Materials* (Thousand Oaks, CA: SAGE Publications, 1998).

Denzin, Norman K. and Yvonna S. Lincoln (eds), *The Landscape of Qualitative Research* (Thousand Oaks, CA: SAGE Publications, 1998).

Diener, Alexander C., "Review of *Seeking Asylum: Human Smuggling and Bureaucracy at the Border*, by Alison Mountz," *The Professional Geographer* 63, no. 4 (2011): 563–565.

Dittmer, Jason, "Review of *Liquid Landscape: Geography and Settlement at the Edge of Early America*, by Michele Currie Navakas," *Social & Cultural Geography* 19, no. 6 (2018): 831–832.

Dufty-Jones, Rae and Chris Gibson, "Making Space to Write 'Care-fully': Engaged Responses to the Institutional Politics of Research Writing," *Progress in Human Geography* 46, no. 2 (2022): 339–358.

Finlay, Linda, "How to Write a Journal Article: Top Tips for the Novice Writer," *European Journal for Qualitative Research in Psychotherapy* 10 (2020): 28–40.

Fiske, Donald W., "Planning and Revising Research Reports," in *Writing and Publishing for Academic Authors*, 2nd edition, edited by Joseph M. Moxley and Todd Taylor (Lanham, MD: Rowman & Littlefield, 1997), 71–82.

Foote, Kenneth E. "Creating a Community of Support for Graduate Students and Early Career Academics," *Journal of Geography in Higher Education* 34, no. 1 (2010): 7–19.

Fregonese, Sara, Diana Martin, and Adam Ramadan, "Commentary: The New Geopolitics of Responsibility in Barack Obama's Cairo Speech," *Environment and Planning D: Society and Space* 27 (2009): 951–955.

Gaile, Gary L. and Cort J. Willmott (eds), *Geography in America at the Dawn of the 21st Century* (Oxford: Oxford University Press, 2003).

Galef, David, *The Supporting Cast: A Study of Flat and Minor Characters* (University Park, PA: The Pennsylvania State University Press, 1993).

Gandy, Matthew, "Books, Geography and Disciplinary Status—An Anglo-American View," in Kevin Ward, Ron Johnston, Keith Richards, Matthew Gandy, Zbigniew Taylor, Anssi Paasi, Roddy Fox, Margarita Serje, Henry Wai-chung Yeung, Trevor Barnes, Alison Blunt, and Linda McDowell, "The Future of Research Monographs," *Progress in Human Geography* 33, no. 1 (2009): 107–109.

Geography Compass, "Aims and Scope," available at https://onlinelibrary.wiley.com/page/journal/17498198/homepage/productinformation.html (accessed April 16, 2021).

Geography Compass home page, available at https://onlinelibrary.wiley.com/page/journal/17498198/homepage/productinformation.html (accessed April 13, 2021).

GeoJournal home page, available at https://www.springer.com/journal/10708 (accessed April 13, 2021).

Geomorphology, available at https://www.sciencedirect.com/journal/geomorphology (accessed April 16, 2021).

Ghertner, D. Asher, *Rule by Aesthetics: World-Cass City Making in Delhi* (Oxford: Oxford University Press, 2015).

Gies, Joseph and Frances Gies, *Life in a Medieval City* (New York: Harper Perennial, 1981 [1969]).

Gillespie, Kathryn A. and Patricia J. Lopez, "Introducing Economies of Death," in *Economies of Death: Economic Logics of Killable Life and Grievable Death*, edited by Patricia J. Lopez and Kathryn A. Gillespie (New York: Routledge, 2015), 1–13.

Griffith, Paddy, *The Viking Art of War* (London: Greenhill Books, 1995).

Guida, Ross J., Scott R. Abella, William J. Smith, Haroon Stephen, and Chris L. Roberts, "Climatic Change and Desert Vegetation Distribution: Assessing Thirty Years of Change in Southern Nevada's Mojave Desert," *The Professional Geographer* 66, no. 2 (2014): 311–322.

Hall, Tim, "Making Their Own Futures? Research Change and Diversity Amongst Contemporary British Human Geographers," *The Geographical Journal* 180, no. 1 (2014): 39–51.

Hamlin, Madeleine, "Second Chances in the Second City: Public Housing and Prisoner Reentry in Chicago," *Environment and Planning D: Society and Space* 38, no. 4 (2020): 587–606.

Harper, Earl T. and Doug Specht (eds), *Imagining Apocalyptic Politics in the Anthropocene* (New York: Routledge, 2022).

Hart, John Fraser, "Ruminations of a Dyspeptic Ex-Editor," *The Professional Geographer* 28, no. 3 (1976): 225 – 232.

Harvey, David, "Editorial: The Geographies of Critical Geography," *Transactions of the Institute of British Geographers* 31 (2006): 409 – 412.

Hawkins, Roberta, Maya Manzi, and Diana Ojeda, "Lives in the Making: Power, Academia and the Everyday," *ACME: An International Journal for Critical Geographies* 13, no. 2 (2014): 328 – 351.

Hetherington, Kevin, "Secondhandedness: Consumption, Disposal, and Absent Presence," *Environment and Planning D: Society and Space* 22 (2004): 157 – 173.

Hood, Ann, *Creating Character Emotions* (Cincinnati, OH: Story Press, 1998).

Hubbard, Phil, Rob Kitchin, Brendan Bartley, and Duncan Fuller, *Thinking Geographically: Space, Theory and Contemporary Human Geography* (London: Continuum, 2002).

Hudson, Peter James, "'The Lost Tribe of a Lost Tribe': Black British Columbia and the Poetics of Space," in *Black Geographies and the Politics of Place*, edited by Katherine McKittrick and Clyde Woods (Cambridge, MA: South End Press, 2007), 154 – 176.

Human Geography, "About This Journal," available at https://journals.sagepub.com/home/hug (accessed April 14, 2021).

Human Geography, "Manuscript Submission Guidelines," available at https://journals.sagepub.com/author-instructions/HUG (accessed April 14, 2021).

Human Geography, "Submission Guidelines", available at https://journals.sagepub.com/author-instructions/HUG#Articletypes (accessed January 23, 2023).

Hunter, Richard, "Land Use Change in New Spain: A Three-Dimensional Historical GIS Analysis," *The Professional Geographer* 66, no. 2 (2014): 260 – 273.

Inwood, Joshua and Derek H. Alderman, "Urban Redevelopment as Soft Memory-Work in Montgomery, Alabama," *Journal of Urban Affairs* (2020), doi: 10.1080/07352166.2020.1718507.

Jakle, John A., "The Writing of Scholarly Books in Geography," in *On Becoming a Professional Geographer*, edited by Martin S. Kenzer (Caldwell, NJ: The Blackburn Press, 2000 [1989]), 124 – 124.

Jang, Woo and Xiaobai Yao, "Tracking Ethnically Dividing Commuting Patterns Over Time: A Case Study of Atlanta," *The Professional Geographer* 66, no. 2 (2014): 274 – 283.

Jarosz, Lucy, "Review of *French Beans and Food Scares: Culture and Commerce in an Anxious Age*, by Susanne Freidberg," *Annals of the Association of American Geographers* 96, no. 1 (2006): 230 – 232.

Jarvis, Helen, "Book Reviews," in *Publishing in Geography: A Guide for New Researchers*, edited by Alison Blunt and Catherine Souch (London: Royal Geographical Society, 2008), 16 – 18.

Jones, John Paul III, Heidi J. Nast, and Susan M. Roberts (eds), *Thresholds in Feminist Geography: Difference, Methodology, Representation* (Lanham, MD: Rowman & Littlefield, 1997).

Kenyon, Sherrilyn, *Everyday Life in the Middle Ages: The British Isles from 500 to 1500* (Cincinnati, OH: Writer's Digest Books, 1995).

Kirsch, Scott and Colin Flint, "Introduction: Reconstruction and the Worlds that War Makes," in *Reconstructing Conflict: Integrating War and Post-War Geographies*, edited by Scott Kirsch and Colin Flint (Burlington, VT: Ashgate, 2011), 3 – 28.

Kirsch, Scott and Colin Flint (eds), *Reconstructing Conflict: Integrating War and Post-War Geographies* (Burlington, VT: Ashgate, 2011).

Kitchen, Rob, "Engaging Publics: Writing as Praxes," *Cultural Geographies* 21, no. 1 (2014): 153 – 157.

Lee, Cameron C., Thomas J. Ballinger, and Natalia A. Domino, "Utilizing Map Pattern Classification and Surface Weather Typing to Relate Climate to the Air Quality Index in Cleveland, Ohio," *Atmospheric Environment* 63 (2012): 50–59.

Legg, Stephen (ed.), *Spatiality, Sovereignty and Carl Schmitt: Geographies of the Nomos* (New York: Routledge, 2011).

Li, Wei, "Chinese Americans: Community Formation in Time and Space," in *Contemporary Ethnic Geographies in America*, edited by Ines M. Miyares and Christopher A. Airriess (Lanham, MD: Rowman & Littlefield, 2007), 213–232.

Li, Wei, Canfei He, and Huaxiong Jiang, "Spatial and Sectoral Patterns of Firm Entry in China," *The Professional Geographer* 71, no. 4 (2019): 703–714.

"List of Journals—Geography," OOIR, available at https://ooir.org/journals.php?category=geography (accessed January 19, 2023).

Liu, Zhiqiang, "Global and Local: Measuring Geographical Concentration of China's Manufacturing Industries," *The Professional Geographer* 66, no. 2 (2014): 284–297.

Lorimer, Jamie, *Wildlife in the Anthropocene: Conservation after Nature* (Minneapolis, MN: University of Minnesota Press, 2015).

Madge, Clare, "On the Creative (Re)turn to Geography: Poetry, Politics and Passion," *Area* 46, no. 2 (2014): 178–185.

Mahtani, Minelle, "Toxic Geographies: Absences in Critical Race Thought and Practice in Social and Cultural Geography," *Social & Cultural Geography* 15, no. 4 (2014): 359–367.

Manzi, Maya, Diana Ojeda, and Roberta Hawkins, "'Enough Wandering Around!': Life Trajectories, Mobility, and Place Making in Neoliberal Academia," *The Professional Geographer* 71, no. 2 (2019): 355–363.

Martin, Ron, "Why Edit Books? In Defense of an Oft-Disparaged Academic Activity," *Regional Studies* 47, no. 9 (2013): 1611–1613.

May, Jon and Paul Cloke, "Home," in *Encyclopedia of Human Geography*, edited by Barney Warf (Thousand Oaks, CA: SAGE Publications, 2006), 225–226.

McCutcheon, Marc, *The Writer's Digest Sourcebook for Building Believable Characters* (Cincinnati, OH: Writer's Digest Books, 1996).

McDonnell, Rachel, "Review of *Watershed*, by Ticky Fullerton," *Progress in Physical Geography* 27, no. 1 (2003): 143–144.

McKittrick, Katherine and Clyde Woods (eds), *Black Geographies and the Politics of Place* (Cambridge, MA: South End Press, 2007).

McKittrick, Katherine and Clyde Woods, "'No One Knows the Mysteries at the Bottom of the Ocean,'" in *Black Geographies and the Politics of Place*, edited by Katherine McKittrick and Clyde Woods (Cambridge, MA: South End Press, 2007), 1–13.

Minca, Claudio and Rory Rowan, *On Schmitt and Space: Interventions* (New York: Routledge, 2016).

Mitchell, Don, "Review of *Writing Worlds: Discourse, Text and Metaphor in the Representation of Landscape*, edited by Trevor Barnes and James S. Duncan," *The Professional Geographer* 45, no. 4 (1993): 474–475.

Mitchell, Don, "Sticks and Stones: The Work of Landscape (A Reply to Judy Walton's 'How Real(ist) Can You Get?')" *The Professional Geographer* 48, no. 1 (1996): 94–96.

Mitchell, Don, "Book Review: *Rebel Cities: From the Right to the City to the Urban Revolution*," *Human Geography* 7, no. 1 (2014): 117–120.

Miyares, Ines M. and Christopher A. Airriess (eds), *Contemporary Ethnic Geographies in America* (Lanham, MD: Rowman & Littlefield, 2007).

Monk, Janice, "Review of *Villages, Women and the Success of Dairy Cooperatives in India: Making Place for Rural Development,* by Pratyusha Basu," *The Professional Geographer* 64, no. 3 (2012): 464–465.

Moseley, William G., "Engaging the Public Imagination: Geographers in the Op-Ed Pages," *The Geographical Review* 100, no. 1 (2010): 109–121.

Mountz, Alison, "State," in *Encyclopedia of Human Geography,* edited by Barney Warf (Thousand Oaks, CA: SAGE Publications, 2006), 460–462.

Mountz, Alison, "Women on the Edge: Workplace Stress at Universities in North America," *The Canadian Geographer/Le Géographe canadien* 60, no. 2 (2016): 205–218.

Mountz, Alison, Anne Bonds, Becky Mansfield, Jenna Loyd, Jennifer Hyndman, and Margaret Walton-Roberts, "For Slow Scholarship: A Feminist Politics of Resistance through Collective Action in the Neoliberal University," *ACME: An International Journal for Critical Geographies* 14, no. 4 (2015): 1235–1259.

Munro-Stasiuk, Mandy J., Timothy G. Fisher, and Christopher R. Nitzsche, "The Origin of the Western Lake Erie Grooves, Ohio: Implications for Reconstructing the Subglacial Hydrology of the Great Lakes Sector of the Laurentide Ice Sheet," *Quaternary Science Reviews* 24 (2005): 2392–2409.

Muzaini, Hamzah and Claudio Minca (eds), *After Heritage: Critical Perspectives on Heritage from Below* (Northampton, MA: Edward Elgar, 2018).

Nash, Catherine, *Genetic Geographies: The Trouble with Ancestry* (Minneapolis, MN: University of Minnesota Press, 2015).

Nast, Heidi J. and Steve Pile (eds), *Places Through the Body* (New York: Routledge, 1998).

Nelson, Jake R. and Tony H. Grubesic, "Oil Spill Modeling: Mapping the Knowledge Domain," *Progress in Physical Geography* 44, no. 1 (2020): 120–136.

Nevins, Joseph, "Academic Jet-Setting in a Time of Climate Destabilization: Ecological Privilege and Professional Geographic Travel," *The Professional Geographer* 66, no. 2 (2014): 298–310.

Obeng-Odoom, Franklin, "Why Write Book Reviews?" *Australian Universities' Review* 56, no. 1 (2014): 78–82.

Oinas, Päivi and Samuli Leppälä, "Views on Book Reviews," *Regional Studies* 47, no. 10 (2013): 1785–1789.

Ortegren, Jason T., Ashley Weatherall, and Justin T. Maxwell, "Subregionalization of Low-Frequency Summer Drought Variability in the Southeastern United States," *The Professional Geographer* 66, no. 2 (2014): 323–332.

Peet, Richard, "Review of *The City as Text: The Politics of Landscape Interpretation in the Kandyan Kingdom,* by James S. Duncan," *Annals of the Association of American Geographers* 83, no. 1 (1993): 184–187.

Peet, Richard, "Discursive Idealism in the 'Landscape-as-Text' School," *The Professional Geographer* 48, no. 1 (1996): 96–98.

Price, Marie, "Andean South Americans and Cultural Networks," in *Contemporary Ethnic Geographies in America,* edited by Ines M. Miyares and Christopher A. Airriess (Lanham, MD: Rowman & Littlefield, 2007), 191–211.

Progress in Human Geography, "Journal Description," available at https://journals.sagepub.com/description/PHG (accessed April 16, 2021).

Progress in Physical Geography, "Aims and Scope," available at https://journals.sagepub.com/aims-scope/PPG (accessed April 16, 2021).

Puāwai Collective, "Assembling Disruptive Practice in the Neoliberal University: An Ethics of Care," *Geografiska Annaler: Series B, Human Geography* 101, no. 1 (2019): 33–43.

Pulido, Laura, "Reflections on a White Discipline," *The Professional Geographer* 54, no. 1 (2002): 42–49.

Pulido, Laura, "Geographies of Race and Ethnicity II: Environmental Racism, Racial Capitalism and State-Sanctioned Violence," *Progress in Human Geography* 41, no. 4 (2017): 524–533.

Richards, Keith, Mike Batty, Kevin Edwards, Allan Findlay, Giles Foody, Lynne Frostick, Kelvyn Jones, Roger Lee, David Livingstone, and Terry Marsden, "The Nature of Publishing and Assessment in Geography and Environmental Studies: Evidence from the Research Assessment Exercise 2008," *Area* 41, no. 3 (2009): 231–243.

Rossi, Ugo and Barney Warf, "Research Monographs and the Making of a Postdisciplinary Geography," *Dialogues in Human Geography* 1, no. 1 (2011): 103–104.

Sear, David, "Review of *Fluvial Geomorphology of Great Britain*, edited by Ken Gregory," *Transactions of the Institute of British Geographers* 24, no. 1 (1999): 125–126.

Shabazz, Rashad, *Spatializing Blackness: Architectures of Confinement and Black Masculinity in Chicago* (Urbana, IL: University of Illinois Press, 2015).

Sheridan, Scott C. and Cameron C. Lee, "Temporal Trends in Absolute and Relative Extreme Temperature Events Across North America," *Journal of Geophysical Research: Atmospheres* 123 (2018): 11,889–11,898.

Skelton, Tracey, "Review of *Cultural Geography*, by Mike Crang," *Area* 31, no. 2 (1999): 190–191.

Smith, Richard, "Review of *Cities, Enterprises and Society on the Eve of the 21st Century*, edited by Frank Moulaert and Allen J. Scott," *Area* 31, no. 2 (1999): 189–190.

Solem, Michael N. and Kenneth E. Foote, "Concerns, Attitudes, and Abilities of Early-Career Geography Faculty," *Annals of the Association of American Geographers* 94, no. 4 (2004): 889–912.

Springer, Simon, *Violent Neoliberalism: Development, Discourse, and Dispossession in Cambodia* (New York: Palgrave Macmillan, 2015).

Starkowski, Kristen H., "'Still There': (Dis)engaging with Dickens's Minor Characters," *Novel: A Forum on Fiction* 53, no. 2 (2020): 193–212.

Swain, Dwight V., *Creating Characters: How to Build Story People* (Cincinnati, OH: Writer's Digest Books, 1990).

Swyngedouw, Erik, *Liquid Power: Water and Contested Modernities in Spain, 1898–2010* (Cambridge, MA: The MIT Press, 2015).

The Geographical Journal, "Author Guidelines," available at https://rgs-ibg.onlinelibrary.wiley.com/hub/journal/14754959/homepage/forauthors.html#TYPES (accessed January 23, 2023).

The Professional Geographer, "The Professional Geographers Guidelines for Commentaries," available at https://www.aag.org/professional_geographer_guidelines_for_commentaries-3/ (accessed January 23, 2023).

Thrift, Nigel, "The Future of Geography," *Geoforum* 33 (2002): 291–298.

Todd, James D., "Experiencing and Embodying Anxiety in Spaces of Academia and Social Research," *Gender, Place & Culture* 28, no. 4 (2021): 475–496.

Tretter, Eliot M., *Shadows of a Sunbelt City: The Environment, Racism, and the Knowledge Economy in Austin* (Athens, GA: University of Georgia Press, 2016).

Turabian, Kate L., *A Manual for Writers of Research Papers, Theses, and Dissertations*, 7th edition (Chicago: University of Chicago Press, 2003).

Turner, Billie L. II, "Whether to Publish in Geography Journals," *The Professional Geographer* 40, no. 1 (1988): 15–18.

Tyner, James A., "The Gendering of Philippine International Labor Migration," *The Professional Geographer* 48, no. 5 (1996): 405–416.

Tyner, James A., "Constructing Images, Constructing Policy: The Case of Filipina Migrant Performing Artists," *Gender, Place and Culture* 4, no. 1 (1997): 19–35.

Tyner, James A., "Scaled Sexuality and the Migration of Filipina Overseas Contract Workers," *Philippine Population Review* 1, no. 1 (2002): 103–123.

Tyner, James A., *Made in the Philippines: Gendered Discourses and the Making of Migrants* (New York: Routledge Curzon, 2004).

Tyner, James A., *The Geography of Malcolm X: Black Radicalism and the Remaking of American Space* (New York: Routledge, 2006).

Tyner, James A., "Filipinos: The Invisible Ethnic Community," in *Contemporary Ethnic Geographies in America*, edited by Ines M. Miyares and Christopher A. Airriess (Lanham, MD: Rowman & Littlefield, 2007), 251–270.

Tyner, James A., *Landscape, Memory, and Post-Violence in Cambodia* (Lanham, MD: Rowman & Littlefield, 2017).

Tyner, James A., "Gender and Sexual Violence, Forced Marriages, and Primitive Accumulation during the Cambodian Genocide, 1975–1979," *Gender, Place & Culture* 25, no. 9 (2018): 1305–1321.

Tyner, James A., "Genocide," in *The International Encyclopedia of Geography*, edited by Douglas Richardson, Noel Castree, Michael F. Goodchild, Audrey Kobayashi, Weidong Liu, and Richard Marston (New York: John Wiley & Sons, 2021).

Tyner, James A., *Red Harvests: Agrarian Capitalism and Genocide in Democratic Kampuchea* (Morgantown, WV: West Virginia University Press, 2021).

Tyner, James A. and Stian Rice, "Along Came a Spider ... and Capitalism Killed It," *Human Geography* (2022), https://doi.org/10.1177/19427786221140843 (accessed January 23, 2023).

Tyner, James A., Mandy Munro-Stasiuk, Corrine Coakley, Sokvisal Kimsroy, and Stian Rice, "Khmer Rouge Irrigation Schemes during the Cambodian Genocide," *Genocide Studies International* 12, no. 1 (2018): 103–119.

Velednitsky, Stepha, Sara N.S. Hughes, and Rhys Machold, "Political Geographical Perspectives on Settler Colonialism," *Geography Compass* 14, no. 6 (2020): e12490.

Walton, Judy, "Bridging the Divide—A Reply to Mitchell and Peet," *The Professional Geographer* 48, no. 1 (1996): 98–100.

Walton, Judy R., "How Real(ist) Can You Get?" *The Professional Geographer* 47, no. 1 (1995): 61–65.

Ward, Kevin and Jo Bullard, "Publishing Books," in *Publishing in Geography: A Guide for New Researchers*, edited by Alison Blunt and Catherine Souch (London: Royal Geographical Society, 2008), 28–38.

Webster, Natasha and Meighan Boyd, "Exploring the Importance of Inter-departmental Women's Friendship in Geography as Resistance in the Neoliberal Academy," *Geografiska Annaler: Series B, Human Geography* 101, no. 1 (2019): 44–55.

Webster, Natasha A. and Martina Angela Caretta, "Early-Career Women in Geography: Practical Pathways to Advancement in the Neoliberal University," *Geografiska Annaler: Series B, Human Geography* 101, no. 1 (2019): 1–6.

Weil, Ben H., "Standards for Writing Abstracts," *Journal of the American Society for Information Science* September–October (1970): 351–357.

Wheeler, James O., "Writing Abstracts," *Urban Geographer* 17, no. 4 (1996): 283–285.

Williams, Miriam, "Review of *A Bun in the Oven: How the Food and Birth Movements Resist Industrialization*, by Barbara Katz Rothman," *Gender, Place & Culture* 24, no. 12 (2017): 1807–1808.

Williamson, J.N. (ed.), *How to Write Tales of Horror, Fantasy, and Science Fiction* (Cincinnati, OH: Writer's Digest Books, 1987).

Wilson, Robert M., "Writing Geography: Teaching Research Writing and Storytelling in the Discipline," *EPA: Economy and Space* 54, no. 7 (2022): 1450–1459.

Witlox, Frank, "Getting Your Paper Reviewed and Finally Published in Journal of Transport Geography: The Do's and Don'ts from the Viewpoint of the Editor-in-Chief," *Journal of Transport Geography* 81 (2019), doi.org/10.1016/j.trangeo.2019.102545.

Wood, Lydia, L. Kate Swanson, and Donald E. Colley III, "Tenets for a Radical Care Ethics in Geography," *ACME: An International Journal for Critical Geographies* 19, no. 2 (2020): 424–447.

Wood, Monica, *Description* (Cincinnati, OH: Writer's Digest Books, 1995).

Woods, Clyde, "'Sittin' on Top of the World': The Challenges of Blues and Hip Hop Geography," in *Black Geographies and the Politics of Place*, edited by Katherine McKittrick and Clyde Woods (Cambridge, MA: South End Press, 2007), 46–81.

Yeung, Henry Wai-chung, *Strategic Coupling: East Asian Industrial Transformation in the New Global Economy* (Ithaca, NY: Cornell University Press, 2016).

Zielke, Julia, Matthew Thompson, and Paul Hepburn, "On the (Im)possibilities of Being a Good Enough Researcher at a Neoliberal University," *Area* (2022), doi: 10.1111/area.12815.

Index

Note: References followed by "n" refer to notes.

https://doi.org/10.1515/9783111189727-012